American Forts

AMERICAN FORTS

Architectural Form and Function

Willard B. Robinson

Published for the
AMON CARTER MUSEUM OF WESTERN ART, FORT WORTH
by the
UNIVERSITY OF ILLINOIS PRESS
Urbana Chicago London

The Amon Carter Museum was established in 1961 under the will of the late Amon G. Carter for the study and documentation of westering North America. The program of the Museum, expressed in publications, exhibitions, and permanent collections, reflects many aspects of American culture, both historic and contemporary.

LIBRARY OF CONGRESS CATALOGING IN PUBLICATION DATA

Robinson, Willard Bethurem, 1935–
    American forts—architectural form and function.

    Bibliography: p.
    Includes index.
    1. Fortification—United States—History.
    2. Military architecture—United States—History.
    I. Amon Carter Museum of Western Art, Fort Worth, Tex.
    II. Title.
    UG410.R6   725'.18'0973   76-25130
    ISBN 0-252-00589-9

DEDICATED TO MY FAMILY

Contents

Illustrations

Maps

Acknowledgments

That the efforts of the scholar are greeted with so much interest and enthusiasm from virtually every individual and agency to which he turns is indeed a fine commentary on society today. At home and abroad, many have generously provided information and material for this work.

First, I wish to acknowledge the authors of the idea. To Mitchell A. Wilder, Director of the Amon Carter Museum of Western Art, belongs the credit for originating this study on forts; I deeply appreciate being given the opportunity by Mr. Wilder to make this study. To Philip Johnson, architect, the Amon Carter Museum and I are indebted for the concept of pursuing the subject from an architectural point of view.

The entire staff at the Amon Carter Museum was exceptionally helpful throughout the duration of my association with them. I particularly appreciate the assistance of Annequinn Allison, former Librarian, and Nancy G. Wynne, Librarian, in locating books and periodicals relative to the subject. In organizing the work and planning research, it was a pleasure to have the counsel of Barbara Tyler, formerly Curator of History with the museum, but now Chief Curator, Museum of Man, Ottawa. Her knowledge of sources was vital. In the final phases of the development of this study it was a pleasure to work with Ronnie C. Tyler, Curator of History of the Amon Carter Museum, and Margaret McLean, former newspaper archivist with the museum.

The staffs of the individual departments of many archives and museums also provided kind assistance. In Canada, the Manuscript Division, Historical Photographs Section, and Map Division provided much guidance in locating material in the Public Archives. The staffs of the Old Military Records Division, Still Pictures Branch, and Cartographic Branch of the National Archives and Records Service in Washington, D.C., spent many hours locating material and fulfilling my requests. A. P. Muntz, Chief of the Cartographic Branch, was especially helpful. I am grateful to the Archivo General de Indias, Seville; the Servicio Histórico Militar, Madrid; the Public Records Office, London; the Map Room, British Museum, London; and the Bibliothèque Nationale, Paris. I am especially indebted to Marie Antoinette Menier, Section Outre-mer of the Archives Nationales, Paris, for assistance in locating manuscript drawings of French colonial works. J. George Stewart, Architect of the Capitol, Washington, D.C., assisted in the location of paintings of fortifications that belong to the Capitol collection.

The United States National Park Service, which has restored and maintains many excellent historic structures, provided photographs and documented material. I appreciate the help of Jim Massey, former Chief, Historic American Buildings Survey, in answering requests for information and graphic material. In addition, I wish to acknowledge the help of Jim C. Gott, Historian, Fort Caroline National Monument; W. H. Glover, Superintendent, Fort Frederica National Monument; Paul C. Swartz, Superintendent, Fort Sumter National Monument; and Pat H. Miller, Chief Park Naturalist, Fort Jefferson National Monument.

I likewise received generous assistance from state and local institutions. My sincere appreciation extends to Edward P. Hamilton, Director of Fort Ticonderoga; Grove McClellan, Publicity Director, Old Fort Niagara Association; Wallace F. Workmaster, Curator of History, Fort Ontario; Nathaniel N. Shipton, Map and Print Librarian, William L. Clements Library, University of Michigan; Henry Grunder, Curator of the Virginia Collection, Earl Gregg Swem Library, The College of William and Mary in Virginia; Mary B. Prior, Secretary, The South Carolina Historical Society; Martha B. Burns, Huguenot Society of South Carolina; James L. Heslin, Director, New York Historical Society; Lee L. Burtis, Head, Iconography Department, California Historical Society; Glenn B. Skillin, Director and Librarian, Maine Historical Society; Myron Leslie Hurwitz, Information Officer, The Port of New York Authority; and the staff of the Illinois State Historical Society, none of whom I have met but all of whom were generous with information.

Eleanor S. Brockenbrough, Assistant Director, Confederate Museum, provided help with paintings of Civil War subjects. Special thanks are due Jeannette D. Black, Curator of Maps at the John Carter Brown Library, who generously assisted me with research in a fine collection of early maps, drawings, and books. At West Point, New York, Joseph M. O'Donnell, Chief, Archives and History Section, United States Military Academy Library, provided many leads, and Gerald Stowe, Curator of History, guided me to defensive works in the vicinity.

I also appreciate the efforts of the staffs of the Special Documents and Interlibrary Loans sections of the Texas Tech University Library in locating and obtaining documents relating to fortifications. For his cooperation throughout the duration of work on this project, my gratitude extends to Nolan E. Barrick, **Chairman**, Department of Architecture, Texas Tech University.

To my cousin Gertrude L. Smith, one of my mentors for many years, my special thanks. Last, but certainly most important, is my family, who patiently traveled with me. Without Peg, my wife, my task would have been tremendously more difficult. She helped with photography, typing, and virtually every other phase of the work.

Conventions Followed in This Work

For consistency in dating and identification, it has been advantageous to establish several conventions for this work. To indicate the chronological development of military architecture, the dates given in parentheses with each structure identify the year that construction was commenced. Two dates separated by a dash indicate the year that erection began and the year that it was completed. However, since modifications and repairs were constantly being made, it is difficult, in many cases, to establish accurate terminal points. Dates of completion—and of occupation—generally have been considered of minor importance since they have little significance in the creative conception and development of architectural types. There are examples of forts that were destroyed or abandoned, after which new works were commenced on or near the same site. Dates pertaining to each phase are separated by a comma. The date of the illustration appears after its identification. Illustrations have been chosen from artists' and engineers' renderings; all have been checked by comparison with extant work and/or written descriptions.

In many cases, the renaming of forts tends to create confusion. To avoid inconsistencies between the text and early drawings and documents, the original name is used while the most recent identification appears in parentheses where the work is first mentioned in the book.

Authorship of plans also creates some confusion since several engineers often contributed to the design of a single fort. Insofar as possible, the artist who most significantly influenced the design and construction is acknowledged. For the western forts of the United States, this person was the officer in charge of the establishment of the post.

Finally, it should be noted that this study covers only military forts. Such independent works as fur trader posts and civilian forts—worthy of extensive studies themselves—are beyond the scope of this book.

American Forts

Fortification, or Military Architecture,
is no other thing than an Art, which
teaches Men to Fortifie themselves with
Ramparts, Parapets, Moats, Covert Ways
and *Glacis's*, to the end the Enemy may
not be able to attack such a part without
great loss of his Men; and that the small
Number of Soldiers which defend the Place
may be able to hold out for some time.

Vauban

Introduction

Historical Development
of the Architectural Theory of Defense
in Europe and Western Asia

Throughout the centuries of its history, the Western Hemisphere has been the scene of innumerable conflicts. Men of this section of the world seem to have been seldom satisfied with the status quo and have aggressively sought new horizons, regardless of the cost to themselves or their neighbors. Inevitably, objectives resulting from the unrest of some peoples have become incompatible with the objectives of others. Consequently, whether motivated by avarice or by conflicting religious or political ideals, the inclination to wage war has characterized most European and, later, American history. Warlike ambitions, whether justifiable or not, have proved to be prominent attributes of Western man.

Societies desiring peaceful ways have often been compelled to defend their right to live tranquilly. In the past, as now, it has seldom been possible to remain completely aloof from those with differing points of view. Even the most peace-loving societies have had to defend their lives, liberty, and fortunes and have thereby rediscovered, from time to time, the futility of attempting to remain neutral without some method of assuring independence. Therefore, from the earliest times, man has contrived structures to provide his physical safety. According to Henri Pirenne, prominent Belgian

historian: "War is as old as humanity, and the construction of fortresses almost as old as war. The first buildings erected by man seem, indeed, to have been protecting walls."[1] No architectural type appears to have had a longer history than fortification.

ANCIENT WORKS FOR DEFENSE

The arms race, too, is as old as mankind. Developments in defense have always followed advances in warfare tactics and new weapons. Throughout many centuries, man has schemed to develop new maneuvers and instruments of combat which would give him advantages over his enemies, but each new offensive weapon and tactic has been answered with innovations for defense. Better methods have been quickly copied and often improved by rivals.

The form and arrangement of the architectural components of defenses have invariably been based upon anticipated modes of siege. Ages ago man discovered the strength-in-numbers concept: several men armed with primitive weapons could easily overcome an individual. For purposes of mutual defense, individuals banded together, formed communities,

3

and surrounded themselves with diverse forms of enclosure (earthworks, rough stone walls, stakes planted in the ground) that made access to the interior difficult except through a controlled passage or gate. These same walls also often provided defenders with advantageous heights from which they could hurl missiles at their besiegers.

In ancient times nearly every important community was fortified in some manner. Archeological research has revealed that cities of antiquity in the Near East were surrounded by massive walls with towers of brick and stone. Jerusalem was among the strongest fortresses built before Christ, occupying the summit of a hill and surrounded by the valleys of the Kedron and the Hinnom, both of which were deep with precipitous sides. Numerous formidable walls and towers were constructed, enlarged, and reconstructed during the many periods of that city's history, and at the time it was razed by Titus (A.D. 70), it consisted of three sections: the Upper City surrounded by the first walls of David; the Lower City enclosed by the second walls built by Jotham, Hezekiah, and Manasseh; and the New City defended by the third, or outer, wall erected by Herod. According to the Jewish historian Flavius Josephus (37–ca. 100), the number of towers in the first wall was sixty, the second forty, and the third ninety-nine; the circumference of the city was nearly four miles.[2] The role of these and other defenses in the history of Jerusalem was told by no fewer than seventeen sieges in fifteen centuries; twice the walls were destroyed.

The particular configurations of ancient fortifications in West Asia were developed in response to the pressures of attack. Walls were made very thick for structural stability and for resistance to the destructive action of pedreros (engines that threw stones) and battering rams, some of which were immense. To further limit the effectiveness of besiegers wielding rams and scaling ladders and to prevent access to the base of the walls, moats were constructed. Thus strengthened, such military architecture could be breached with the weapons then in use only by vastly superior numbers of men.

GREEK AND ROMAN FORTIFICATIONS

Fortification was likewise a well-developed art of the ancient Greeks and Romans. In Greek cities, defense was beautifully integrated with other functions. Prominent topographical features were selected for the dignified locations of altars, statues, treasuries, and temples—a consideration which was also compatible with defense. By surrounding these irregular hilltops with walls, citadels were created that not only protected life but also provided security for the most important buildings. The Acropolis in ancient Athens, like many other Greek sacred precincts, was surrounded by a very high wall and was accessible only through a single gate. The city that surrounded this famous hill was, like the hill, enclosed by thick masonry walls.

Among the ancient Romans, necessity continued the practice of walling cities, but the destructive power of machines of war had been well advanced. Marcus Vitruvius Pollio (first century B.C.), a Roman architect, engineer, and author, described ballistae (engines resembling crossbows) capable of throwing some missiles weighing up to 360 pounds.[3] He also described the ram-tortoise (enclosed battering ram), a contrivance consisting of a timber framework from which a heavy, rope-suspended ram could be swung. The framework was mounted on a platform with wheels and had a roof covered with rawhide or other material to protect the besieging soldiers from the solid and burning missiles hurled by the defenders. Similar roofed devices were used to protect soldiers while they filled moats with stones and earth so that they could approach the walls with battering rams or other devices.

Mining was another offensive technique used to attack walls. In this operation, attackers starting from a safe distance away dug tunnels toward the fort until they reached the wall. Then the subterranean works were continued directly under the wall for some distance. As the work progressed, the tunnels were propped and shored with dry wooden members—material often obtained from nearby buildings. When the undermining of the defensive walls was complete, the tunnel was filled with combustible material, the props were greased, and the whole was burned, whereupon it and the walls of the fort above it collapsed. Mining was often accomplished in spite of the efforts of the defenders to countermine, or dig tunnels of their own to intersect the underground works of the besiegers, and to drive them out by combat or with smoke from burning faggots, resin, and tar.

The potential destruction of such siege operations created a need for defenses of greater and greater strength. To enclose their towns, the Romans constructed "two strong walls of masonry, separated by an interval of twenty feet: the space between was filled with earth from the ditches, and loose rock well rammed, forming at top a parapet-walk, slightly inclined towards the town to allow the water to pass off: the outer of these two walls, which was raised above the parapet-walk, was massive and crenellated; the inner one was very slightly elevated above the ground level of the place inside, so as to render the ramparts easy of access."[4] Vitruvius wrote that the city walls should be "laid out not as an exact square nor with salient angles, but in circular form, to give a side view of the enemy from many points" and should be strengthened by "towers . . . set at intervals of not more than a bowshot apart."[5]

The Roman military genius was adept not only in permanent fortifications, but in the construction of temporary works as well. A considerable part of the success of the Romans in extending their vast empire was in their proficiency at military engineering in the field. Caesar, for example, was a master of this art. In enemy territory he fortified every camp whether it was intended to be occupied for only a day or for an entire season. Consistent in spirit with the discipline of the Roman military organization, camps were exactly quadrangular in form unless need dictated otherwise. "As soon as the space was marked out, the pioneers carefully levelled the ground, and removed every impediment that might interrupt its perfect regularity."[6] According to Gibbon, "The camp of a Roman legion presented the appearance of a fortified city."[7]

Prior to the fall of the Roman Empire, many European towns were walled by the Romans as defense against barbarians. For several centuries after, few developments occurred in the art of fortification in Europe. In England, France, and Italy, under the direction of the church, old Roman fortifications were repaired and used as the need arose.

Medieval Military Architecture

In the Byzantine Empire fortification was a well-developed art from the fifth through the thirteenth centuries. Heavy walls, sometimes double and concentric, reinforced by carefully placed high towers and deep moats, surrounded cities such as Constantinople. Walls were battlemented, and gateways were protected with flanking towers. While the degree of influence these fortifications had on later developments is now difficult to ascertain, it is certain that the Crusaders carried this Asiatic style back to Europe.

In the ninth century new structures for defense—baronial and royal—appeared in France, built through fear of raids from neighboring peoples. These early forts or castles, simple and of impermanent construction, usually consisted of wooden towers or keeps placed upon natural or artificial hillocks or mounds. Built in connection with the keeps were stockades or palisades, for which the labor was supplied by peasants. The enclosure was usually circular in plan and contained within, in addition to the keep, churches, quarters, workshops, and magazines. The mound was commonly surrounded by another palisade and a ditch which was crossed at the gate by a drawbridge. Within this outer ring of defenses there was room to shelter peasants and stock when raids occurred. By the eleventh century this French type of fortification, termed a "motte and bailey," had spread to England and to other countries of Europe.

While the early mottes and baileys were easy to erect from earth and wood, they obviously were neither durable nor long defensible against strong attacks. They decayed rapidly, were easily burned by besiegers, and could be readily destroyed by engines of war such as the ram and the catapult. Therefore, stone soon replaced wood in the construction of walls and keeps.

When stone construction first appeared, castle walls were simple curtains without towers or other defensive adjuncts. With time, however, battlements were developed, and towers were added as the heights increased. The plain, straight wall, experience proved, was difficult to defend; when attempting to inhibit the enemy's destructive operations at the base of the wall, defenders were forced to expose themselves as they looked down its face. Therefore, the practice was developed of constructing hoardings, or wooden galleries, cantilevered several feet over the edge of the wall. From openings in the hoarding floor, stones and boiling water could be employed effectively against besiegers at the foot of the wall. Later, in the thirteenth century, stone corbeling replaced the impermanent and flammable wood hoardings, and the familiar projecting battlemented wall, with its merlons and crenels, evolved.

Early in the age of chivalry, the custom of flanking curtain walls with lofty towers projecting beyond the face of the enceinte was adopted to enable defenders to keep siege engines away from the walls. In keeping with earlier Roman and Byzantine practices, these towers were located at salients and, in long walls, at intermediate positions within bow-shot of each other so that two towers would enfilade each curtain. Because of the inherent strength of cylindrical forms and their better resistance to blows of missiles and battering rams, the towers were usually round, were small in diameter, and had no more than two or three loopholes at each story looking along the wall. By the end of the twelfth century, the use of towers of a variety of sizes and configurations to flank both sides of gates had become prevalent.

During the thirteenth century, to further increase their strength, castles were planned so that although some sections of the fortifications might fall, others would remain defensible. Towers, particularly at the gate, were constructed to be held even if the enemy breached the wall. This new concept led to the development of the concentric castle, wherein lines of defense were successively developed within each other. When an outer line fell to a besieging force, the castle was still tenable from the next line inward; in fact, when the besieged left a defense they often retreated to a stronger position. Natural barriers such as water, cliffs, and ravines were often incorporated into the system as a part of one of the lines.

The Renaissance of the Art of Fortification

In Europe, then, there was considerable progress in the art of fortification during medieval times, the highest degree of perfection being attained in the castles of France, which were almost impregnable. During this period, for defensive advantages, many strongholds were placed on rocky eminences. The desire for height, by similar reasoning, was also apparent in the design of the various architectural forms incorporated into the castles of the kings and barons. Walls, towers, and gatehouses were all increased in height and thickness to give the besieged the advantage of elevation for their defensive weapons while making more difficult the forceful entry by besiegers, in response to medieval siege tactics and the engines of war then in use.

But new developments in weapons led to a renaissance in the art of fortification. Methods of attack improved so radically that within a relatively short time a revolution was created in architecture for defense. Gunpowder, first invented by the ancient Chinese, was known in medieval Europe. However, early firearms were crude and small and were of little concern to castle builders. Bombards, predecessors of cannons, were cone-shaped devices which fired rough stone missiles. These weapons, without carriages, were fired as they rested directly on the earth; they were awkward, heavy, and inaccurate. Although they were ineffective at first, cannons firing stone projectiles were used as early as 1340 against fortress walls.[8]

At the end of the fourteenth century, methods of producing cannons and gunpowder had been developed sufficiently to make them generally useful in siege operations. Early cast iron or cast bronze models fired lead or iron balls and, like the bombards, rested directly on the earth. The effectiveness of cannons was further increased when they became lighter and when carriages to move them efficiently were devised.

By the fifteenth century, artillery had the power and accuracy to pulverize stone walls. Castle walls and towers, looming at great heights to resist escalade, were perfect targets. In 1494 Charles VIII and his French troops stunned Italy when they marched through the country, reducing castle after castle with their amazing bronze artillery. Attack became superior to defense.

The immediate reaction to the violent destruction inflicted by artillery was the modification of the high, medieval architecture. Adjustments were sought to incorporate defensive artillery, to alter existing fortifications, and to design new ones capable of resisting the new siege techniques. Medieval walls—only approximately nine feet wide—were too narrow and the towers were too confined to mount artillery. Consequently, walls became wider. Projecting towers, still necessary to flank the walls to prevent escalade, were increased in breadth and were open at the back to provide free access for men and guns. Lower parts of towers were sometimes provided with embrasures to enfilade the curtains, while upper parts were designed to allow cannons to fire outward. Other modifications were made in defensive enclosures. To reduce the amount of surface exposed to cannon fire and to make them more stable, walls and towers were made lower and were supported from behind with masses of soft earth. To stabilize and increase the resilience of walls when struck by missiles, wooden members were often interwoven into them— a practice advocated by Vitruvius in Roman times. The construction of stone merlons was eventually discontinued; when struck by cannonballs they shattered into fragments which did more killing than the missiles themselves.[9]

ICHNOGRAPHY OF BASTIONED FORTIFICATION

Among the very serious inadequacies of medieval fortifications was the relatively poor flanking arrangement of the walls and towers. To effectively resist a strong besieging force, it was necessary that the defensive architecture allow for good protection of all parts of the enceinte and nearby ground. The small, round towers of medieval times had insufficient capacity to allow defenders to concentrate heavy fire along the curtains and ditch, and, when towers were increased in size, there was a corresponding increase in the amount of dead area—area which could not be covered from some other part of the strong-hold. In this undefended space at the foot of the enceinte, a besieger was relatively free to attack the walls by mining or escalade, hindered only by missiles which could be thrown over the edge. To eliminate the dead ground created by the curvature of a round or nearly round form, the projecting features were given a geometrical configuration that con-formed to the outline of the dead area—the undefended por-tion was included in the enceinte. Works were projected farther from the curtains to provide good coverage of the faces and to create better conditions for enfilade fire from the flank. The projecting form which eventually resulted was the bas-tion.

Appropriately enough, the innovation of the bastion oc-curred in Italy. The inventor of the plan may have been Francesco di Giorgio Martini late in the fifteenth century, but the first to build one was probably Michele Sanmicheli (1484–1559), early in the sixteenth century.[10] In search of effective configurations, engineers gave bastions many forms in the early years. At first they were small; in plan, some were nearly round with only a slight point, some resembled elon-gated circles, and others were spadelike in appearance. In time, the areas increased, and finally it was found that the simple salient was the most effective shape.

In establishing the plan of a bastioned fortification, it was mandatory that there be no area of land that could not be covered by musketry; every part of the fort had to be seen and defended from some other part. Thus, the flanks of the bas-tions were the most critical and important parts of the en-ceinte, since they contained the arms that enfiladed the oppo-site faces of the bastion, reverse-fired on the opposite flanks, and, with the guns at the opposite end of the front, crossfired on the ditches in front of the curtains. If they failed to fulfill their function of denying the enemy access to the enceinte, the main body of the place was greatly exposed to capture by escalade or assault. Like bastion faces, flanks evolved through many designs.

For effective flanking fire, most engineers maintained that the distance from the flank to the point of the opposite bastion (the length of the line of defense) could not exceed the effective range of musketry, which at the beginning of the seventeenth century was approximately 200 to 240 yards but which had increased to nearly 310 yards by the early eighteenth century. This distance influenced the number of bastions in a wall. A square, bastioned form was the most basic trace for a defensible enceinte.[11] However, the small flanks and the sharp flanked angles inherent in the geometry of this configuration cramped key interior spaces and made it more difficult to defend than works on polygons with more sides. Moreover, since it could enclose only a small area without exceeding the permissible length for lines of defense, the square, bastioned trace was useful only for citadels, detached works, or small wilderness forts. As the need for interior space grew, the length of enclos-ing walls became greater, increasing the number of bastions. Thus, the number of sides on the exterior polygon was corre-spondingly increased, which resulted in pentagonal, or hexa-gonal, bastioned enceintes, enceintes based on other superior polygons, and irregular polygons. Since the area of towns to be enclosed was large, superior polygons were generally re-quired for them.

Although most theories for tracing a bastioned fort were idealistically based on regular geometry, the surrounding terrain—hills, swamps, water, expected approaches for attack, and so on—often necessitated numerous adjustments in de-sign. In order to conform to the terrain, many forts were constructed with traces developed from irregular polygons.

Once the enceinte was determined, the fortification could be further strengthened by the incorporation of various outworks (defenses located between the enceinte and the glacis). The size and character of these outworks depended upon the nature of the ground, anticipated methods of attack, and economy. In form and placement they always conformed to rather definitive principles and harmonized geometrically with the configuration of the enceinte.

A ditch invariably surrounded the main body of the fort. Whether wet or dry, its main function was to force aggressors into an exposed position in front of the enceinte. When water was available, a wet ditch was often integrated into the bastioned system of defense, particularly when the stronghold was garrisoned by a relatively small force. As in the medieval period, surrounding the enceinte with water countered mining and presented a formidable obstacle to an enemy intent upon storming the place. However, a wet ditch handicapped communication between the main body of the fort and its outworks and had the additional disadvantage of restricting sorties by the garrison. For large garrisons, consequently, the dry ditch was often favored by some engineers.

The outworks could be expanded by the inclusion of a ravelin (a work comprised of two faces and two demigorges). Located in front of a curtain, it functioned to protect that curtain and the adjoining bastion flanks from enemy cannonading. It was used most often in front of sallyports and was separated from the main body of the fort and other outworks by a ditch.

It was discovered, probably in the sixteenth century, that the defensive strength of a fortification could be increased by inclusion of a covered way. Placed on the outer side of the ditch, the covered way served as a protected road of communication around the fort. It often had a parapet and a banquette from which grazing fire could be brought upon the glacis to hinder the enemy from gaining acess to the ditch. To prevent the possibility of besiegers occupying the covered way and establishing breeching batteries there, it was completely covered from the main body of the fort—consistent with the maxim, "What offers itself as a defence ought to be defended."[12]

By rounding the counterscarp of the covered way, salient places of arms were formed; by introducing an angular projection from the covered way at reentering angles, reentering places of arms were provided. At these locations, additional batteries could be mounted and sorties could be organized.

ORTHOGRAPHY OF RENAISSANCE FORTIFICATION

While on the one hand the Renaissance military architect was improving the defensibility of fortifications with bastions and outworks that made approach difficult for besiegers, he was occupied on the other hand with strengthening the works against artillery. In response to the power of siege cannons, theory changed radically concerning the form of fortifications in profile. Instead of building exposed targets high above the terrain, architects soon learned the advantages of excavating and sinking fortifications into the ground. Although walls of the main body of a fort often continued to be of masonry and of sufficient height to prevent escalade, they ideally were protected by massive earthworks.

As in tracing, profiling was based on rather definite principles. Earthworks, in respect to outlying areas, sloped upward toward the main body of the fort. Theoretically, if a permanent fortification were well designed, the besieging force should have been able to see only a series of earth inclines rising one above the other. The glacis sloped up and terminated against a parapet or, if a covered way was not provided, against the counterscarp. Across and above the ditch, as seen from the field, were the exterior and superior slopes of the parapet of the main body of the fort. Thus designed, it was difficult for enemy cannon fire to score hits on masonry. In order to breach the protected ramparts of the fort, it was necessary, under the counterfire of the besieged, to advance batteries close enough to the main body to be able to fire directly at the scarp.

Scarps and counterscarps were usually battered—sloped inward and upward. Statically, this battering provided a stable construction that would resist the pressures of the earth as it tended to push the walls out and would produce walls that were more capable of withstanding the impact of cannon fire.

From within the enceinte and outworks it was mandatory that all inclines and parapets be so profiled that the outlying ground could be completely subjected to grazing musket or cannon fire. Likewise, from within, it was mandatory that the outworks, as well as the outlying ground, be so designed that they could be completely commanded from the main body of the fort lest they fall into enemy hands and be used in the attack.

Finally, minor refinements were made to deter an attacking enemy. To eliminate any objects behind which the enemy might seek cover, trees, rocks, and other variations in terrain were removed. The outlying area was completely cleared and smoothed for a distance at least equal to cannon range. Then, after the main part of the fortification was completed, obstacles designed to entangle the enemy were prepared. Stockades were sometimes constructed in the ditch, and fraises were occasionally set up. On the outlying ground, abatis and chevaux-de-frise were other obstacles that were sometimes employed.

Thus, the horizontal and vertical characteristics of all the architectural elements for defense developed from their function as individual units and from their interdependent function as a combined group of elements. In a well-conceived fortification, all the parts looked like a unit and worked together because of their geometrical and functional relationships—the result of scientific and logical conception.

SYSTEMATIZATION OF THE ART OF FORTIFICATION

During the Renaissance period science and logic led to systematization, that is, the methodical organization of ensembles of components. In a spirit typical of the epoch, these components were formed and arranged into fronts of fortifications according to maxims, which were formulated but which might vary in relative importance when interpreted by different artists. Although each element of the system might exist and function as an entity, if well designed, the enceinte and outworks gained strength by their interdependence.

Named after their individual originators, after general geometrical characteristics, or, when adopted officially, after a country, many systems of fortification were developed. Diverse forms and types of architectural elements were combined with different enceinte traces having varying degrees of complexity. The most complicated arrangements were generally the strongest against attack, but they were also the most expensive to build and defend.

In addition to the bastioned system preferred by the French, Dutch, and Italians and the polygonal system preferred by the Germans, the tenaille system was occasionally used in the sixteenth century. In this arrangement, the enceinte was composed simply of a series of salient and reentrant angles. Although engineers who favored this system believed that it was advantageous because of its long flanks, many other theorists observed that it created dead ground at the reentrant angles which could not be adequately defended without elaborate outworks. Moreover, since the development of satisfactory flanking conditions required reentering angles of not less than sixty degrees and a many-sided regular polygon of defense, it was limited in flexibility. It appears, therefore, to have been rarely used after the sixteenth century, although advocates appeared afterward.[13]

The polygonal system was admired for its simplicity, economy, and adaptability to both large and small enclosures. Main enclosures were traced in the forms of simple polygons, without reenterings, and therefore contained, per unit length of wall, more area than the bastioned or tenaille configurations. Instead of providing for flanking fire from the enceinte itself, caponniers were constructed in the main ditch to enfilade the scarp. These caponniers were located on short fronts in the center of the sides and were spaced apart on long fronts at distances which would allow effective coverage of the entire curtain between them with musketry. Shielded by the glacis, caponniers were better protected than the flanks of the bastioned system. With coverfaces, ravelins, and orillons to shield them, fire for close defense could be continued from them into late periods of a siege. The caponniers themselves were defended from the ramparts.

SÉBASTIEN LE PRESTRE VAUBAN

All of these systems and many variations of them were certainly known and studied by Sébastien Le Prestre Vauban (1632–1707),[14] who carried the art of fortification to new levels of effectiveness and strength while, at the same time, developing new and successful siege tactics. Thirty-three new fortresses were constructed, and almost three hundred others were improved, by this French genius who fortified every frontier town in France.[15] In addition, he directed some fifty sieges.[16]

Although his early work was similar in form to that of his mentor, Blaize-Françoise, Comte de Pagan (1604–65), the illustrious Vauban scientifically improved his designs through a wide range of experience in constructing fortifications, directing their active defense, and serving as engineer on siege operations. Vauban arrived at certain principles, or maxims, which were considered universal in application. First, it was mandatory that all parts of the fortification be viewed, or flanked, from some other part so that there would be no sheltered place where an enemy could lurk. To further the effectiveness of the defenses by providing good coverage of the parts with flanking fire, he admonished that wide flanks were best. Then, every flanked part must be within effective musket range. On the function and construction of the components of a fort, he observed that whatever encloses a durable fortification must either flank, face, or curtain, and that it should be built sturdily enough to withstand the first discharges of cannon directed against it. These were basic determinants of form for his fortifications.[17]

By observing weaknesses in his own work and that of others, Vauban developed three basic systems which have been mentioned in most, if not all, treatises written since his time. The first was the most widely used. It was essentially a simple, straightforward bastioned arrangement with tenailles before the curtains. His second system consisted of an arrangement wherein the bastions were separated from the enceinte and the angles of the latter were reinforced with small, pentagonal works called tower bastions. Theoretically, this separation would allow a main bastion to be destroyed without a breach being made into the main body of the place. The third system, similar to the second, was a more complicated assemblage having added outworks, larger towers, and a more complicated enceinte which was traced to provide, in effect, a second set of flanks. However, the last two systems were not often used.[18]

But Vauban's genius in the art of fortification lay not in systematization per se, but rather in the creative and intuitive, yet scientific, approach he developed for the design of each fort; he was not "one of those *esprits routiniers*."[19] Each stronghold was planned according to its particular circumstances in such a manner that none of the basic principles was violated. Although he held as one of his principles that regular fortification—that is, work of geometrically ordered components—is much to be preferred to irregular,[20] he recognized the difficulty of adapting geometric uniformity to uneven terrain. In the design of new fortifications or the improvement of old ones he studied in great detail all aspects of the water and ground, including the nature of the earth which would be used in the ramparts and on the outworks, before he devised a concept of defense. He was a master at manipulating earth and water to compensate for weaknesses and to capitalize on advantages provided by the terrain. Existing fortifications were also considered and used wherever possible. Because of these considerations, each fortification by Vauban was unlike any that had preceded it.

The perfection of Vauban's style was apparent in the design of Dunkerque, Flanders, which is considered to have been one of his finest works and, with the exception of the fortifications of Louis XIV's Paris, his most famous work. The city was protected by a formidable arrangement of bastions, curtains, ditches, ravelins, tenailles, redoubts, and hornworks. On the west side were *criques* (a series of intersecting wet ditches) designed to impede approach from that direction. The combination of these components made Dunkerque one of the strongest cities in Europe.

In the art of fortification, then, Vauban demonstrated the logic and prowess upon which the strength of military architecture greatly depended. The universality of his principles is indicated by the fact that they remained valid for over a century and a half. Developments in fortifications which occurred after the death of this military genius responded to progress in weapon design and occurred as a result of the introduction of new architectural forms better able to satisfy the essential requirements he had known, not to satisfy entirely new principles.

Vauban had profound influence on both the theory and the practice of the art of defense in Europe and in the New World. In 1690, as a consequence of his influence, a corps of engineers was founded in France. Ultimately, this corps would be significant in America by providing experience with military architecture to men who would serve in the colonies as well as in the United States. On theory of defense, Vauban's first system formed the foundation of academic work in military architecture in France. The highly esteemed École Polytechnique made the precepts of this system the basis for its program on field and permanent fortification and attack and defense of fortified places. Eventually, the United States Military Academy at West Point, founded in 1802, developed a curriculum similar to that of the esteemed French school. Thus, European influence, particularly that of France, was ingrained into the art of defense in America from early colonial times through the nineteenth-century development of the country.

Colonial Fortifications

During the fifteenth and sixteenth centuries, several European nations, awakening to the potential of explorations in unknown lands, turned their attention to voyages westward. It was not, however, until the middle of the sixteenth century that any of these nations attempted to claim any section of the North American continent by settlement. France undertook colonization on the Saint Lawrence River as early as 1541; in the 1560s both France and Spain started colonization on the southeastern Atlantic seacoast; later in the century England attempted to colonize the central coastal region.

Motivation to colonize North America derived from the desire for political freedom, material profit, acquisition of territory, or missionary zeal. Disagreement and competition over these objectives made conflict an inevitable part of colonial life and settlement. Consequently, for protection from natives and from other Europeans, fort building was a mandatory defensive measure from the very beginning.

In colonial America political and physical geography were integrally related to defense. Spain and France both sought to maintain early land claims with forts that were strategically located to limit the expansion of other nations. Since communication was vital to colonization, strongholds were positioned to control harbors and natural interior highways. Forts were built at the mouths of rivers which emptied into the Gulf of Mexico and the Atlantic Ocean; inland, other defensive works appeared on large lakes and at points of confluence of major rivers. Thus, with the maintenance of comparatively few fortifications it was theoretically possible to control large sections of land.

The selection of appropriate locations for fortifications and the design and realization of works to occupy them required the services of individuals knowledgeable in the principles of the art of fortification. As the intensity of competition among nations became greater and as warfare became more methodical, the demand for skilled professional military engineers increased.

EARLY FORTS ON THE FLORIDA COAST

During the latter half of the sixteenth century, characteristic technical skill in the selection of defensible positions and the design of suitable works was exhibited by French military leaders in establishing claims to various sections of the continent. For Fort Caroline, Florida (1564), a site was selected on a

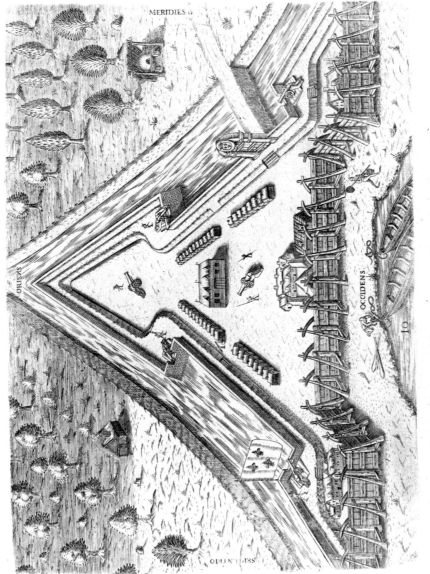

1. Fort Caroline, Florida (1564). René de Laudonnière, engineer. Engraving by Theodore de Bry from a painting by Jacques Le Moyne. *Library of Congress, Washington, D.C.*

bluff overlooking the Saint Johns River at a location approximately five miles from the ocean (fig. 1). At that point the river was comparatively narrow, thus allowing the use of artillery for defense against passing boats.

The fort that was laid out on this site was triangular, with bastions at the angles. Its designer, René de Laudonniere, described the plan and construction of the work:

Our Fort was built in forme of a triangle. The side toward the West, which was toward the lande, was inclosed with a little trench and raised with turves made in forme of a battlement of nine foote high: the other side which was toward the River, was inclosed with a Pallisado of plankes of timber after the manner that Gabions are made. On the South side there was a kind of bastion within which I caused an house for the munition to be built: it was all builded with fagots and sand, saving about two or three foot high with turfes: whereof the battlements were made.... Loe here in briefe the description of our Fourtresse, which I named Caroline in honour of our Prince King Charles.[1]

Upon learning of French activity on the Saint Johns River, Spain sent an expedition to capture Fort Caroline, to establish an outpost in Florida to guard Spanish shipping lanes, which passed along the eastern Florida coast to and from South America, and to drive out other settlers. Under the leadership of Pedro Menendez de Aviles (1519–74), who took Fort Caroline, the first of nine wooden forts was constructed at Saint Augustine.

Like those established elsewhere along the coast during the latter part of the sixteenth and the early part of the seventeenth centuries, the forts at Saint Augustine were simple, short-lasting structures with palisade or stockade enclosures of logs set vertically into the ground—forms of enclosure that were centuries old.[2] One of these forts, the second to be built in 1566, evidently consisted of an area enclosed within a wooden palisade.[3] Walter Biggs, a member of Sir Francis

Drake's expedition, which sacked Saint Augustine in 1586, described another one of these forts, evidently the sixth, Fort Juan de Pinos, as being

built all of timber, the walles being none other but whole Mastes or bodies of trees set up right and close together in maner of a pale, without any ditch as yet made, but wholy intended with some more time; for they had not as yet finished al their worke, having begunne the same some three or foure moneths before: so as, to say the trueth, they had no reason to keepe it, being subiect both to fire, and easie assault.

The platforme whereon the ordinance lay, was whole bodies of long pine trees, whereof there is great plentie, layd a crosse one on another, and some little earth amongst.[4]

Thus built, the forts at Saint Augustine were naturally very impermanent and unstable, for the wood used in them decayed rapidly in the hot, humid climate, requiring frequent repairs.[5] In 1595 it was noted concerning one of the forts (fig. 2), probably by Hernando de Mestas, that "All of it is in danger of collapsing, inside and out.... The guns on its walls are not to be fired because it is feared all the curtains will tumble down."[6] Moreover, as the history of the forts at Saint Augustine bears evidence, if they did not collapse they were subject to burning. Governor Andres Rodriguez de Villegas observed that the timber from which the eighth fort was constructed was "so dry and ready to burn, that merely by using the artillery therein fire breaks out in many places."[7] If Spain was to maintain her foothold in Florida, more permanent works were required.

SPANISH FORTIFICATIONS IN FLORIDA

By the last part of the sixteenth and first half of the seventeenth centuries Spain had recognized the necessity of maintaining possession of the Florida coast for the protection of her

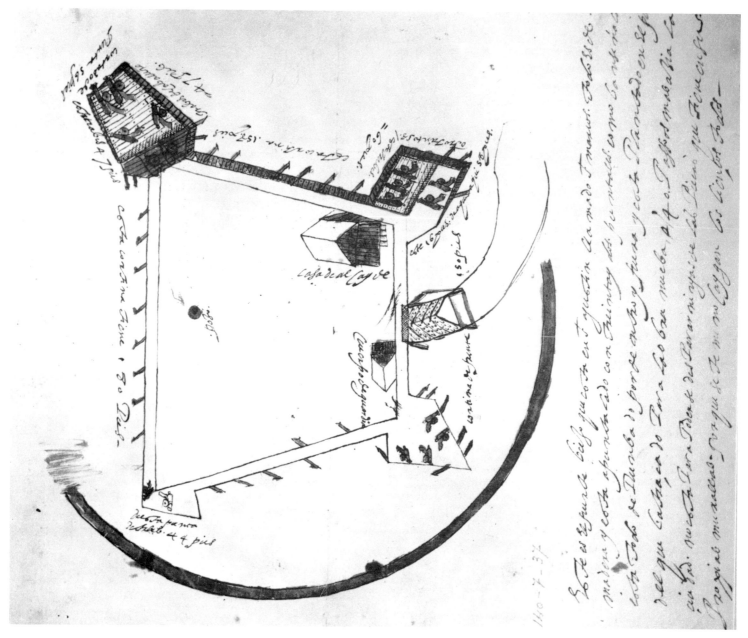

2. Fuerte Biejo que Está en San Augustin, Florida (ca. 1593). Plan (ca. 1593). *Archivo General de Indias, Seville.*

shipping lanes. Yet this very coastal area had remained but weakly and impermanently fortified. The wooden structures for defense at Saint Augustine and at the Matanzas inlet, as history shows, were incapable of resisting pirate raids and could never have withstood military assault.

With Spanish strength and influence in North America declining, and with the continuing threat of pirate raids and of English expansion, the erection of permanent fortifications was authorized in 1669 by an initial appropriation of twelve thousand pesos.[8] After some preliminary preparation, construction was begun in 1672 on the Castillo de San Marcos at Saint Augustine, the center of the Spanish system of defense in Florida (fig. 3).

Using the labor of Indians, slaves, convicts, and free Spaniards—skilled and unskilled—the construction of the Castillo continued for more than eighty years. Evidently designed by Ignacio Daza of Havana,[9] it was first completed in 1696,[10] after which it was unsuccessfully besieged by the English in 1702. Later, to strengthen the work, a covered way and other outworks were added. In 1738 the razing of older rooms in the court and the construction of casemates were started. This remodeling was finally completed in 1756 at a total cost of "30 million dollars, a sum which is said to have caused the king of Spain to remark that 'its curtains and bastions must be made of solid silver.'"[11]

In plan, the Castillo was designed according to established formulas (fig. 4). Architecturally similar to, but smaller than, other forts in Spanish possessions along the Gulf of Mexico and in Central America, it was a nearly square, four-bastioned fort. Characteristically, a wet ditch surrounded the enceinte, and there was a ravelin to defend the gate, which was secured by a drawbridge and portcullis (fig. 5). At each of three bastion salients was located a small bartizan, or lookout; on the fourth or San Carlos Bastion (the northeast bastion) was placed a large watchtower to oversee Matanzas Bay. Although they appeared on late Spanish castles and on later Spanish and French forts in other geographical areas, these cylindrical compartments were not common in North American fortifications.

On the interior, the vaulted casemates opened on all four sides to the parade, which was approximately one hundred feet square. A significant feature—one which reflects in part the Spanish philosophy and mission in the New World—was the strong visual emphasis on the chapel by its axial placement with the sallyport. Storerooms, barracks, magazines, officers' quarters, a guardroom, and a prison occupied other casemates. At the southeast corner of the parade a ramp was located for moving artillery to and from the terreplein.

Functioning as a defense adjunct to Castillo de San Marcos was a system of outer lines begun in 1706. The first of these was evidently "a palisaded or stockaded line of pine, with lunettes or angles."[12] Sometime between 1720 and 1730 two additional lines of earthen outworks with palisading encircled the three landward sides of the town. As a result of these additions, Saint Augustine actually became a fortified town, with the Castillo de San Marcos as the citadel.

In addition to the direct route from the ocean through Matanzas Bay, Saint Augustine could be approached from the south by way of the Matanzas River. After studying the defenses of the settlement, Antonio de Arredondo, an engineer from Havana, recommended that a permanent, fortified outpost be erected on the south end of Anastasia Island to watch this approach. There, under the direction of the engineers, Arredondo and Pedro Ruiz de Orlando of Saint Augustine, Fort Matanzas was started in 1736. Six years later, the small fort was completed.

Fort Matanzas was a lucid example of form expressive of function (fig. 6). For good defensive position, the main plat-

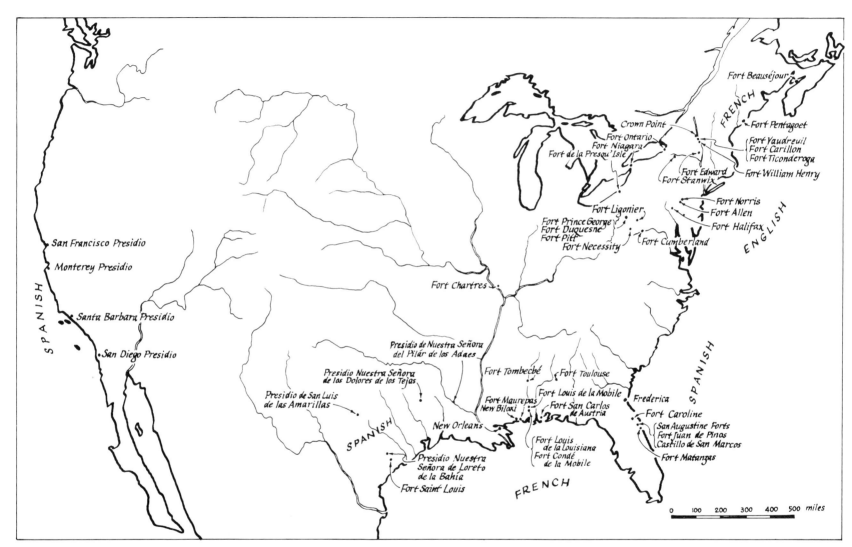

Fort Beauséjour

FRENCH

Crown Point
Fort Ontario
Fort Niagara
Fort de la Presqu'Isle

Fort Pentagoet
Fort Vaudreuil
Fort Carillon
Fort Ticonderoga
Fort William Henry

Fort Edward
Fort Stanwix

Fort Norris
Fort Allen
Fort Halifax

ENGLISH

Fort Ligonier
Fort Prince George
Fort Duquesne
Fort Pitt
Fort Necessity

Fort Cumberland

San Francisco Presidio

Monterey Presidio

Fort Chartres

SPANISH

Santa Barbara Presidio

San Diego Presidio

Presidio de Nuestra Señora
del Pilar de los Adaes

Presidio Nuestra Señora
de las Dolores de los Tejas

SPANISH

Fort Tombecbé
Fort Toulouse

Presidio de San Luis
de las Amarillas

Fort Louis de la Mobile
Fort Maurepas
New Biloxi
Fort San Carlos
de Austria

Frederica

Fort Caroline
San Augustine Forts
Fort Juan de Pinos
Castillo de San Marcos

SPANISH

New Orleans

Fort Louis
de la Louisiana
Fort Condé
de la Mobile

Fort Matanzas

Presidio Nuestra
Señora de Loreto
de la Bahía

Fort Saint Louis

FRENCH

0   100   200   300   400   500   miles

Colonial Fortifications

1770
THE LAST GUESTS.

3. Castillo de San Marcos, Florida (1672–1756). Ignazo Daza and others, engineers. Woodcut. *United States Military Academy, West Point.*

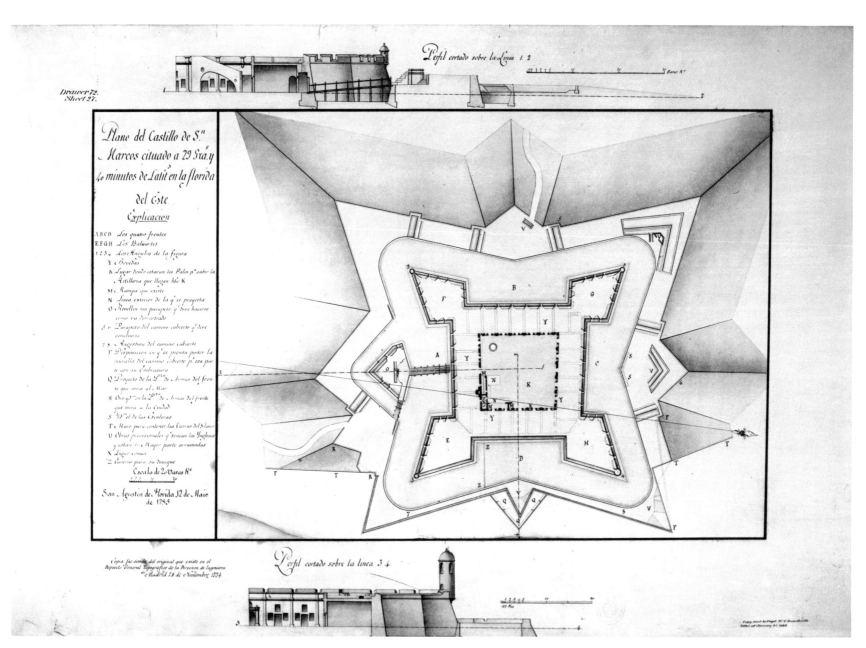

4. Castillo de San Marcos, Florida (1672–1756). Plan (1785); facsimile (1884). *National Archives, Washington, D.C.*

1821
SALUTING THE FLAG.
SPANISH TROOPS
EVACUATING FT. MARION

5. Castillo de San Marcos, Florida (1672–1756). Woodcut. *United States Military Academy, West Point.*

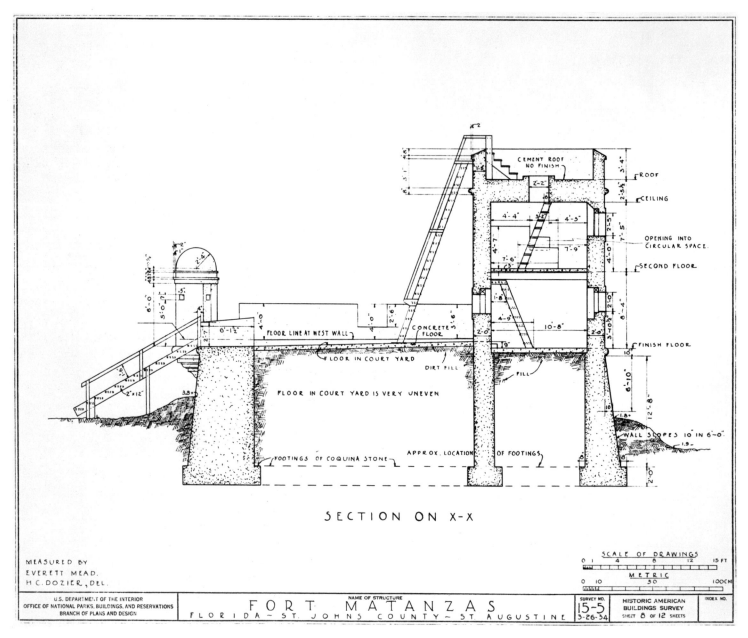

SECTION ON X-X

MEASURED BY
EVERETT MEAD.
H.C. DOZIER, DEL.

| U.S. DEPARTMENT OF THE INTERIOR OFFICE OF NATIONAL PARKS, BUILDINGS, AND RESERVATIONS BRANCH OF PLANS AND DESIGN | NAME OF STRUCTURE FORT MATANZAS FLORIDA ~ ST. JOHNS COUNTY ~ ST. AUGUSTINE | SURVEY NO. 15-5 3-26-34 | HISTORIC AMERICAN BUILDINGS SURVEY SHEET 8 OF 12 SHEETS | INDEX NO. |

6. Fort Matanzas, Florida (1736). Pedro Ruiz de Orlando, engineer. West elevation drawn by George G. Cellar. *Historic American Buildings Survey*.

form was situated approximately ten feet above the plane of the site. Since the primary purpose of the fort was to ward, it was provided with only two embrasures in the parapet. Above the main platform was a two-story tower, about twenty feet high, surmounted with a parapet. The roof of the tower, from which surveillance could be maintained over the surrounding country, was structured with a vault that spanned it longitudinally.

Although completely different in form and function, there were stylistic similarities between the Castillo de San Marcos and Fort Matanzas: both were built from coquina stone finished with stucco; the battered walls of both were terminated with a cordon; both had bartizans which were supported on molded corbeling and roofed with domes. The openings on Fort Matanzas, like those of the Castillo, had molded casings with ears at the corners. These two works were beautiful examples of Spanish fortifications—the finest of the period in America.

Meanwhile, across the peninsula Spain was also struggling to defend a settlement at Pensacola. In theory, Pensacola Bay, landlocked by nature, was easily controlled by forts located at the narrow inlet. On the northwest side, across from Santa Rosa Point, the ground rises steeply, providing a natural advantage for batteries. There, Fort San Carlos de Austria (1698–99) was built. Designed by Jaime Franck, it was constructed from " pine stakes some twelve inches thick... set deeply into the sand in two parallel rows, perhaps six yards apart. Held in place by these sunken footings two rows of pine logs (each about nine yards in length and one foot thick at the base) leaned inward and upward and joined their tips, like the rafters of a gable, some twenty-five feet above the ground. Within the space thus enclosed, triangular in section ... sand was shoveled and poured even to the top."[13] The temporary nature of this method early became apparent to the Spaniards.

At the end of the next year, even before the fort was completed, the logs had begun to rot and the sand had begun to sift out from between them.

As at Saint Augustine and other key coastal cities, additional works—temporary as well as permanent—appeared for the defense of Pensacola's water approach. Military engineers from Spain, England, and the United States later practiced their art there, but the forts they designed belong to other eras.

EARLY FOUR-BASTIONED FORTS

As demonstrated in Florida, the design of forts during the last part of the seventeenth and the first part of the eighteenth centuries was, to a large degree, a matter of adapting standard forms to the terrain in accordance with economic circumstances, building materials, and anticipated modes of attack. Since most of these defenses, particularly those in remote areas, were small, the number of types of efficient plans used was relatively limited.

The form most frequently used in America was the square, four-bastioned plan widely known in Europe and brought across the ocean by the French, Spaniards, and others. While several methods of developing the trace of a fort with this geometrical configuration were precisely formulated, it proved to be a flexible form in the construction that could be used. It could be erected from earth, from logs laid horizontally or set in the ground vertically, or from stone or brick according to required strength, economy, manpower, and the skill of the laborers. Within limits, the size could be varied to accommodate various needs of the garrison.

The French demonstrated this flexibility well at diverse locations. Fort Pentagoet, Maine (1635–45), had an enceinte of earth and stone, the curtains of which were sixteen feet thick at the base and six feet wide at the top (fig. 7).[14] Obviously designed for a small garrison, the parade was only about sixty feet square.

Another fine example of the four-bastioned arrangement was Fort Maurepas (1699) on the Bay of Biloxi, built under the direction of Le Moyne d'Iberville (1661–1706), who had been sent to make secure France's claims to the lower Mississippi Valley (fig. 8). According to a narrative of the expedition, *"The fort was made with four bastions, two of them of squared logs, . . . placed one upon the other, with embrasures for port holes, and a ditch all around. The other two bastions were stockaded with heavy timbers which took four men to lift one of them. Twelve guns were mounted."*[15] The bastions of squared logs (the Royal Bastion and the Bastion of the Chapel) were diagonally opposite and were provided with wooden decks above which rose log parapets.[16] The other two bastions (the Bastion of Biloxi and the Bastion of the Sea) along with the four curtains were formed by double rows of timbers planted vertically in the ground. This heavy construction indicates that the fort was designed to resist the fire of light field artillery. Outside the main body of the fort was a stockade with redans for additional defensive strength.

The Forts Louis in the Mobile Bay area were similar to Fort Maurepas in form and construction. The first, Fort Louis de la Mobile (1702), on Twenty-Seven Mile Bluff on the Mobile River, had four bastions of logs.[17] It was laid out following the plan of Sieur Le Vasseur,[18] and erection was directed by Joseph de Serigny and François Bienville. The second, Fort Louis de la Louisiana (1711), was farther downstream. According to the drawing and description by Sieur Cheuillot, it was a square, bastioned fort of cedar logs which contained the governor's house, magazine, guardhouse, and jail. Both forts formed the nuclei of surrounding settlements.

About 1723, under the direction of Sieur Devin,[19] a cartographer and draftsman, work commenced on a permanent fortification at Mobile which replaced the second wooden structure that decayed so rapidly. The new work, called Fort Condé de la Mobile, was described in 1735 as being built from brick, with four bastions, demilunes, ditch, covered way, and glacis.[20] It was the strongest military work erected by the French on the Gulf Coast. When surveyed early in the nineteenth century, it was found to measure approximately three hundred feet between the salients of the bastions, and it was provided with vaulted bombproof chambers. Within the curtains was a series of casemates, and within the northwest and southeast bastions were a bakery and a powder magazine.[21]

Like Condé, the second Fort Chartres, Illinois (1753–56), was a permanent work on a square trace with bastions at each corner. Designed for Indian defense only, it was about 490 feet on each side and was fortified with stone walls slightly over two feet thick, replacing decaying wooden works set up in 1718. An eighteen-foot-high enclosure and bastions with embrasures, loopholes, and a sentry station provided for an active defensive function.[22] Following the conflict between France and England, it was described by Philip Pittman, a British officer, as the " best built fort in North America."[23]

The square, four-bastioned trace and variations of it continued to be a frequently used form in French Louisiana. Fort Toulouse (1717), built by Le Blond de la Tour (d. 1723) near the confluence of the Coosa and Tallapoosa rivers, was also of logs and had four bastions;[24] Fort Tombecbé, Alabama (1735), in a fine adaption of a regular geometric trace to the topography, had a stockaded enclosure in the form of a modified square with one bastion and two demibastions defending two sides while the other two overlooked the natural obstacles of a ravine and the Mobile River.[25] These forts were both examples of primitive construction adapted to sophisticated plans.

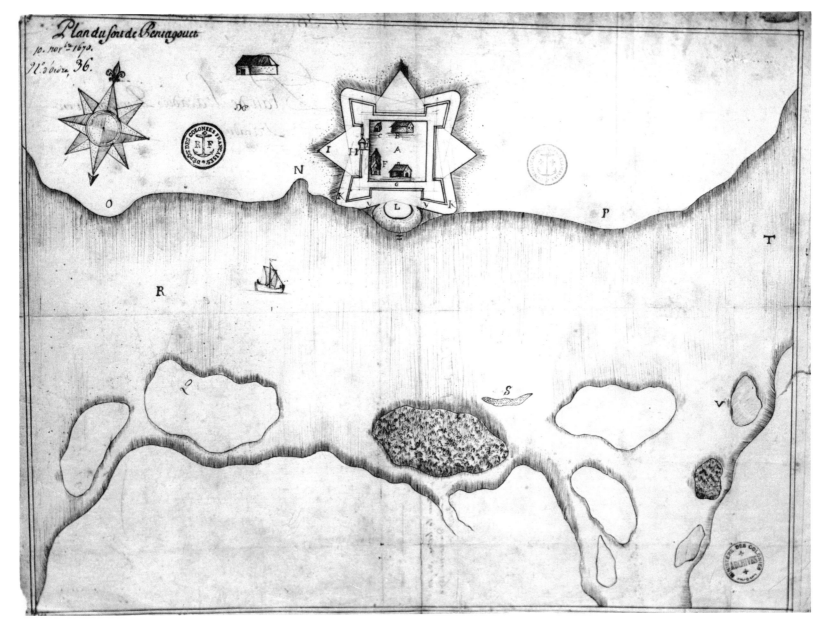

7. Fort Pentagoet, Maine (1635–45). Plan (1670). *Archives Nationales, Paris.*

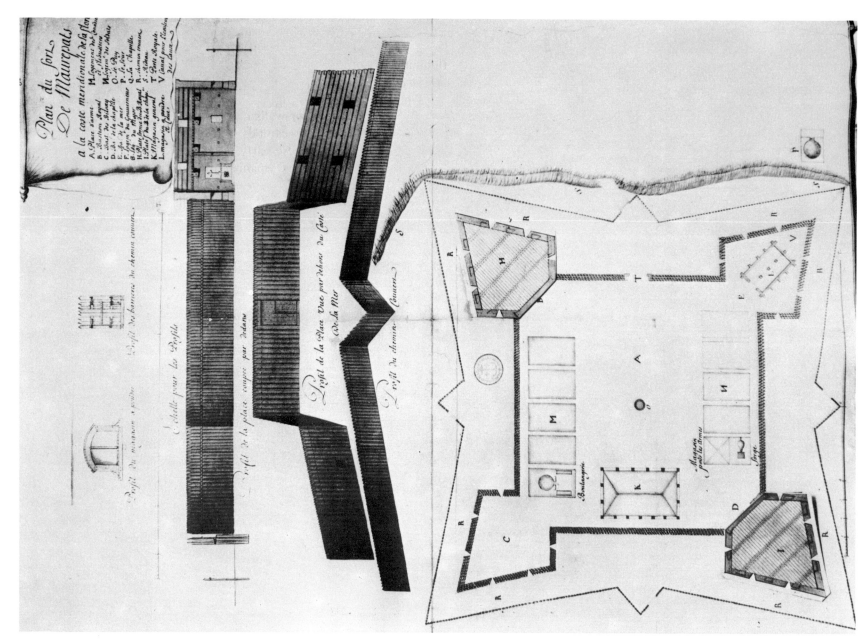

8. Fort Maurepas, Mississippi (1699). Plan and details (1705); facsimile. *Public Archives of Canada, Ottawa.*

PRESIDIOS

Although many of the colonial forts built by Spain and France were small and primitively constructed, they were, nonetheless, comparatively costly in terms of the numbers of men required to build and garrison them. In this respect the close relationship that existed between the crown and the church proved to be a great advantage in Spanish efforts at colonization. With little or no reward, missionaries were willing to endure great hardships and danger to go among the Indians and to Christianize them and subjugate them to the crown, thereby establishing Spain's claim to the fields in which they were successful; frontier defense and liturgy went hand-in-hand.

Missionaries recognized the need to select for their work tribes that were peaceful and sedentary. However, Indians were not always predictable in their congeniality; for the protection of the padres, therefore, presidios to garrison soldiers were necessary adjuncts to the missions. In the Southwest several presidios were set up early in the eighteenth century to defend the missions against hostilities. Among them in Texas were Presidio Nuestra Señora de los Dolores de los Tejas (1716); Presidio de Nuestra Señora del Pilár de los Adaes (1721), Spanish capital of Texas until 1722; and Presidio Nuestra Señora de Loreto de la Bahía (1722), located near the site of La Salle's Fort Saint Louis (1685). The plans for fortifications for these presidios were distinguished by pure geometrical relationships of architectural forms. All possessed similarities in concept of arrangement of defensive components, yet they varied in plan. Dolores had an enclosure based on a square trace with bastions on diagonal corners, while Adaes was hexagonal with bastions on alternate angles.[26] Both had enceintes which were evidently composed of palisades.

Built by Marquis de San Miguel de Aguayo, Nuestra Señora de Loreto was developed on an octagon and also had bastions at every other angle, but, unlike the others, it had works (*lengua de sierpe*) with plans which resembled redans projecting from intermediate points (fig. 9). Barracks were arranged concentrically, with the outer walls forming part of the enceinte. Passages located on the capitals furnished access to each bastion—typical of all these early eighteenth-century Texas presidios. Characteristic of Spanish military planning, the chapel was located on an axis with the main gate.

The Presidio de San Luis de las Amarillas (Presidio San Sabá), Texas (1757, 1763), located on the San Saba River, was originally established on the frontier to protect the ill-fated San Sabá Mission approximately three miles away. At one time one of the largest presidios in the province,[27] it consisted of several primitive buildings surrounded by a log stockade. As if to symbolize Spanish determination to maintain this lonely frontier, stone works replaced the original decayed stockade and wooden buildings in 1761 (fig. 10). Nevertheless, all of this building proved inadequate to resist the pressure of the Indians, who raided the fields and pastures, and in 1768 the presidio was abandoned.

The mission system was also called upon to defend Spanish interests along the coast of Alta, or Upper, California. During the latter part of the eighteenth century, four presidios were established to protect twenty-one missions and three pueblos in a chain extending from San Diego to San Francisco Solana.

Characteristic of virtually all of the earliest works in a wilderness area, the first presidios in California were crudely built. In 1796 an engineer reported that the presidio at San Diego (1769) had "no merit than that the enemy would perhaps be ignorant of their weakness."[28] The first works of the Presidio of Monterey (1770) consisted of temporary huts surrounded by a simple stockade of logs.[29] A group of tawdry buildings arranged around a quadrangle comprised the San Francisco Presidio, established in 1776 (fig. 11). Less than two

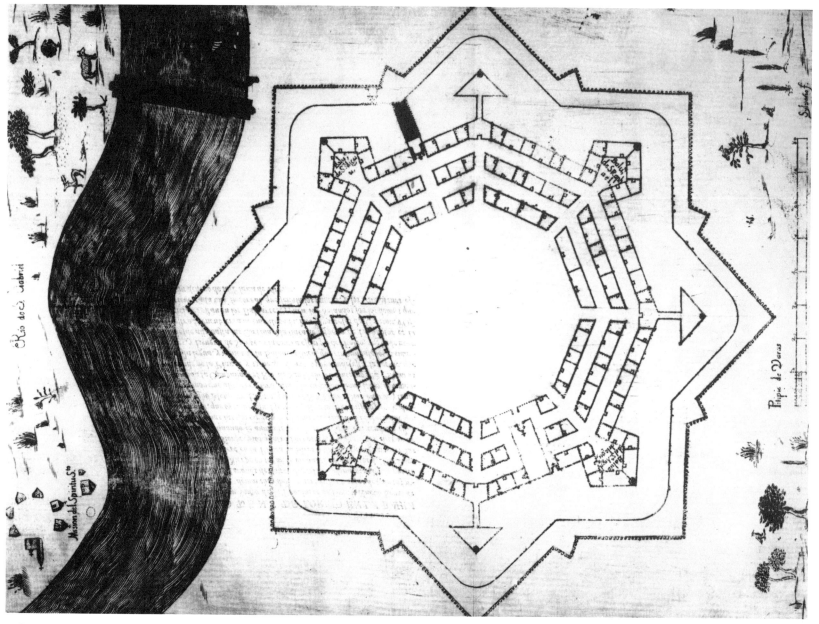

9. Presidio Nuestra Señora de Loreto, Texas (1722). Marquis de San Miguel de Aguayo, engineer. Plan from J. A. de la Pena, *Dorretero de la expedición en la provincia de los Texas . . .* (Mexico, 1722). *John Carter Brown Library, Providence, R.I.*

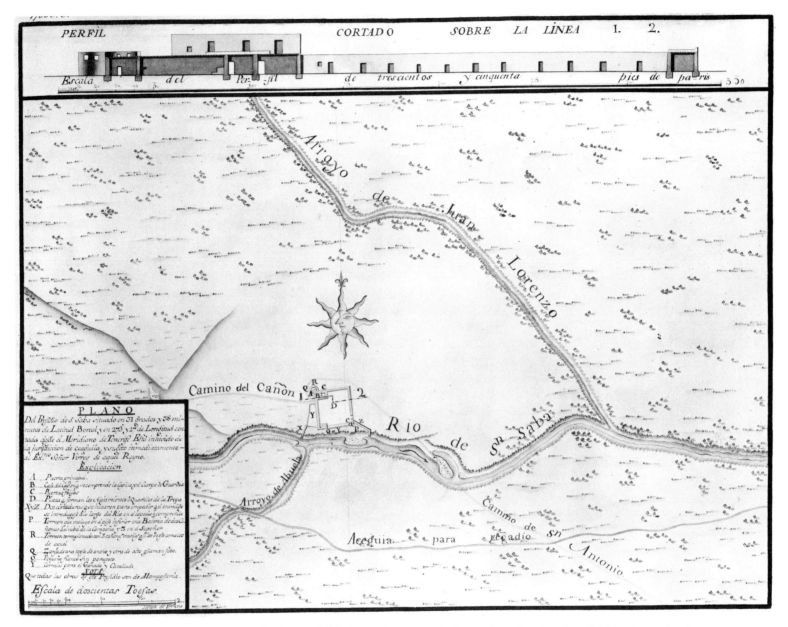

10. Presidio de San Luis de las Amarillas (Presidio de San Sabá), Texas (1757, 1763). Plan and section elevation. *British Museum, London.*

11. San Francisco Presidio, California (1776). View from Louis Choris, *Voyage pittoresque au tour du monde* (Paris, 1822). *Library of Congress, Washington, D.C.*

decades later, this last military work was described by Vancouver, an English traveler, as being "enclosed by a mud wall, and resembling a pound for cattle. Above this wall the thatched roofs or their low houses just make their appearance. One side was very indifferently fenced in by a few brushes here and there, fastened to stakes in the ground. The walls were about fourteen feet high and five feet in breadth, and were formed by uprights and horizontal rafters of large timber, between which dried sods and moistened earth formed into a sort of mud plaster, which gave it an appearance of durability."[30]

Santa Barbara (1782), the fourth to be established, was the finest of the California presidios. In plan it consisted of groups of buildings neatly arranged around a plaza 330 feet square. At diagonal corners were bastions on which cannons were mounted. Located on an elevated part of the plain, it was orderly and well constructed with tile-covered roofs.[31] The outer walls, which enclosed buildings and corrals, were of adobe seven feet thick and were constructed with the assistance of Indian labor.

Thus, in terms of planning and construction the development of presidios was similar to that of forts. In both military types, plans were generally regular and geometrically disciplined, with bastions flanking the walls. However, the earliest structures were crudely and impermanently built. Only with time did the work become more durable—a characteristic that reflected increasing competition over the continent.

French Fortifications at Mid-Century

While Spain was attempting to strengthen her claims with the mission system, France continued to maintain that the key to North America was the natural continental transportation system. Therefore, the French thought, by controlling the waterways with strong forts they could effectively protect their claims. In pursuance of this concept, old fortifications were strengthened and new ones erected at considerable expense. Fort de la Presqu'isle (1753), under the supervision of Chevalier François le Mercier, and Fort de la Riviere au Boeuf (1753), supervised by Pierre-Paul de la Malgue, Sieur de Marin, were intended to control the gateway—Presqu'isle portage—to the Ohio country. Fort Niagara (1718–26) was strengthened, under the direction of François Pouchot between 1750 and 1759, by the addition of new earthworks and batteries. Fort Duquesne (1754) was constructed at the confluence of the Monongahela and Ohio rivers, replacing the captured English Fort Prince George (1754). Fort Carillon (1755), originally called Fort Vaudreuil and known later as the famous Fort Ticonderoga, was built at the south end of Lake Champlain. During the first half of the eighteenth century the French were the most prolific fort builders in colonial America.

In the 1750s France, however, did not possess the manpower or the logistical potential to build elaborate permanent fortifications in the wilderness areas of her claims. Therefore, French forts at that time continued to be small and were erected with expedient constructional methods, although forms were based on scientific design theory.

Unless forced into unusual forms by the terrain, as at Niagara, small forts conformed to the square, four-bastioned trace. But with the French and Indian War (1754–63) at hand, engineers recognized the importance of fortifying against artillery-supported sieges. Therefore, the erection of curtains and bastions demanded structural methods which employed masses of material, since stockades built of single rows of timbers offered virtually no resistance to cannon fire.

A year before Braddock's army was defeated on its way to capture the fort in 1755, Fort Duquesne had been established under the direction of François le Mercier (fig. 12). Like Forts de la Presqu'isle and de la Riviere au Boeuf, erected the previous year, it was a square, bastioned work. Protected from attack by the rivers and a steep incline, the north and west curtains, the northwest bastion, and half of the northeast and southwest bastions were simple, loopholed stockades. Facing the land, the remainder of the curtains and bastions, however, were constructed by erecting two parallel walls, ten and one-half feet apart, consisting of square-hewn logs laid horizontally, one on top of the other.[32] In the process of building, the two walls were tied together by heavy cross-timbers placed at right angles and dovetailed into place. The space between the logs was then filled with earth, thus forming a massive, rather resilient structure. Like the simple stockade, this system of earth and timber fortification had been known in Europe many centuries earlier.[33] Embrasures, also formed with logs, were provided for the cannons at the top of the wall. Such a structure was subject to rapid deterioration, but this problem did not occur at Fort Duquesne, since the French destroyed the fort themselves before retreating in 1759.

Similar to Fort Duquesne in form and construction was Fort Carillon (1755)—renamed Fort Ticonderoga after its capture and reconstruction by the British in 1759. Developed as a square, bastioned fort under the direction of the inexperienced French engineer Marquis Michel Chartier de Lotbinière, who had studied military engineering in France,[34] it was advantageously located on a rocky ridge overlooking Lake Champlain at the outlet of Lake George, with two sides facing the water. On the landward sides the curtains were reinforced with ravelins, since they were most vulnerable to attack.

As at Fort Duquesne, the wooden walls of the main enclosure were double, the space between being filled with earth. But if they had had the resources, de Lotbinière and his men evidently would have built a more permanent work at first.

He wrote: "We were not prepared to build in stone, having neither the material assembled nor the workmen. We were therefore obliged to line the works in oak which fortunately was plentiful on the spot. I began the parapet of the whole work which I formed in a double row of timbers distant ten feet from one another and bound together by two cross-pieces dovetailed at their extremities, to retain the timbers."[35] Later, in 1757, the desire for a durable work led the engineer to order the scarp revetted with a comparatively thin stone veneer. Before the work was completed, however, the French abandoned the fort, leaving it in flames.

Since French forts were located to control passages for transportation and not to take advantage of good defensive positions afforded by natural terrain, they were often weakened by their sizes, and their defensibility was frequently limited by their size. Interestingly, both Forts Duquesne and Carillon possessed these very defects. In the opinion of Louis-Antoine de Bougainville, a French officer, the former was too small and was vulnerable to domination by cannon fire from nearby Mount Washington.[36] The latter, General Marquis de Montcalm thought, should have been twice as large.[37] According to a reconnaissance by the English Lieutenant Dietrich Brehm in 1759, Fort Carillon could also be commanded from nearby Mount Defiance.[38] Later, in the War of Independence, both the British and the Americans capitalized on this drawback.

In addition to Duquesne and Carillon, the French began new fortifications at the mouth of the Niagara River. This site had been used as a location for defensive works since 1678, when La Salle constructed the first crude fort; Marquis de Denonville followed with a four-bastioned stockaded fort in 1687. Designed by Gaspard-Joseph Chaussegros de Léry, engineer-in-chief of New France, the "castle," a thick-walled stone building, was erected in 1726. These early defenses,

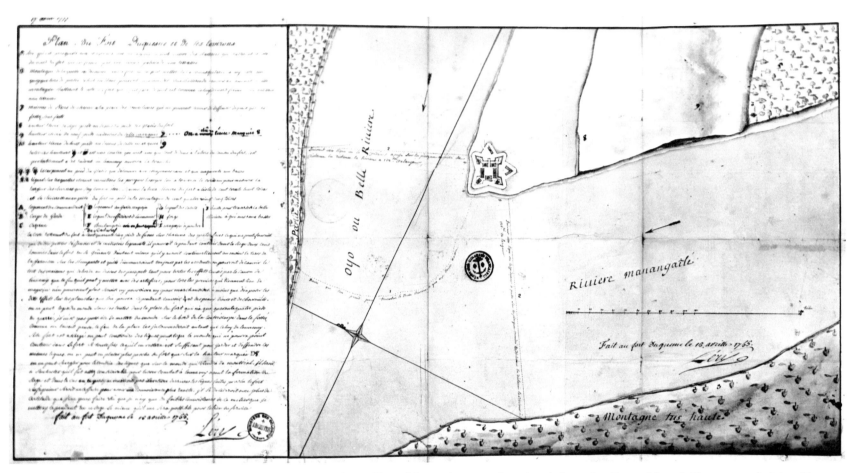

12. Fort Duquesne, Pennsylvania (1754). François le Mercier, engineer. Plan of fort and surroundings (1775) drawn by Gaspard-Joseph Chaussegros de Léry, *fils*. *Archives Nationales, Paris.*

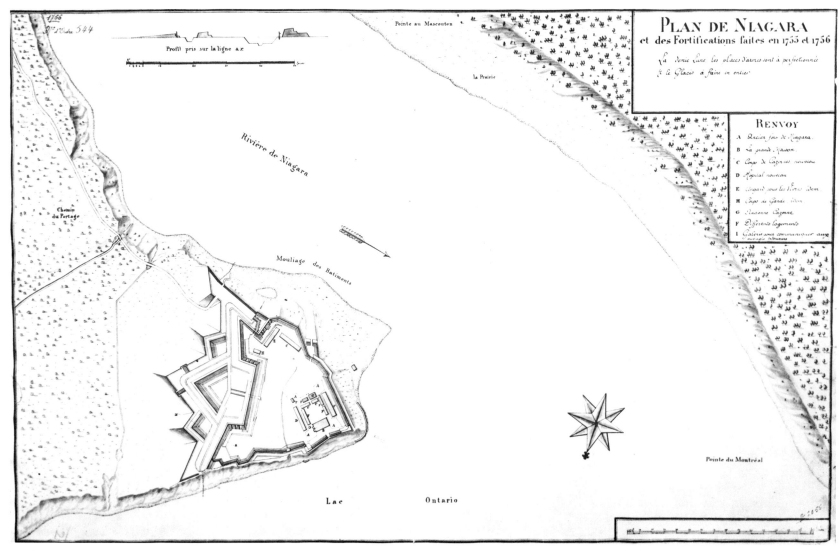

13. Fort Niagara, New York (1679, 1755). François Pouchot, engineer. Plan (1756); fascimile. *Public Archives of Canada, Ottawa.*

however, were incapable of strong resistance against a military siege. Consequently, François Pouchot, royal engineer, was sent to supervise new fortifications.

The irregular trace of the new undertakings at Fort Niagara (1755) demonstrated not only the skill of the designer but also the efficiency and advantages inherent in a carefully selected site (fig. 13). Because of its peninsular location, it was necessary to design against a siege from one direction only. Thus, the earth enceinte defending the land approach was built with modified demibastions and reinforced with a large ravelin, redoubts, and a ditch, all of which were geometrically ordered in plan. However, the sides overlooking the waterway that the fort controlled followed the irregular configuration of the terrain and were not as formidable in mass.

ENGLISH FORTIFICATIONS AT MID-CENTURY

Unlike New France, which had a planned program of fort construction to control English expansion, the British colonies had given virtually no thought to special structures for defense during the early 1750s. In contrast to the French, who brought to the continent a traditional military heritage and a well-organized military system, the English settlers at that time were militarily unorganized, had no skilled engineers, and had little means to construct works for the common defense. There were comparatively few fortifications, and those that existed were in a dismal state of disrepair.

During the middle of the decade, with the approach of the French and Indian War, several new outposts were provided by the Virginia province. However, they were ineffective, barely qualifying as fortifications. For example, Fort Prince George, captured by the French shortly after it was begun in 1754 and replaced by them with Fort Duquesne, was merely a storehouse surrounded by a hastily constructed stockade.

As its name seems to imply, Fort Necessity (1754), rapidly built under the direction of twenty-two-year-old Colonel George Washington (1732–99), consisted of a circular stockade only fifty-three feet in diameter, a small house in the center of the enclosure, and some minor earthworks. A circular trace, while enclosing the greatest possible amount of area for a given perimeter, was indefensible from within, since it was impossible to enfilade the walls.

Late in 1755 the provincial assemblies appropriated some funds to erect fortifications. Most of these authorized forts had small, stockaded enclosures, although there was considerable variation in their traces, depending upon the terrain and the supervisor's judgment. Fort Norris, Pennsylvania (1756), "consisted of a group of buildings surrounded by a stockade eighty-feet square with four, half bastions."[39] Fort Allen, Pennsylvania (1756), built under the supervision of Benjamin Franklin, was rectangular, 125 feet by 50 feet, with two demibastions on diagonal corners and a redan on the middle of each of the two long sides.[40] None of these English fortifications was sufficient to resist military assault or cannon fire; their chief function was the protection of settlements from surprise attacks and raiding parties.

To the northeast, in the vicinity of Lakes George and Champlain, many forts likewise had stockaded enclosures because more formidable works were not possible. Fort Edward, New York (1755), had an enclosure of this type and was in the "form of a square with three bastions." At the time, Captain Eyre, the designer, expressed his opinion of the strength of the work when he wrote that "3 or 400 men will be able to resist 1500, provided they do their duty, if cannon is not brought against it."[41]

As with the French and Spanish, the rectangular, bastioned trace was the form most used by the British provincials for their small forts. Fort Cumberland (1754), the building of

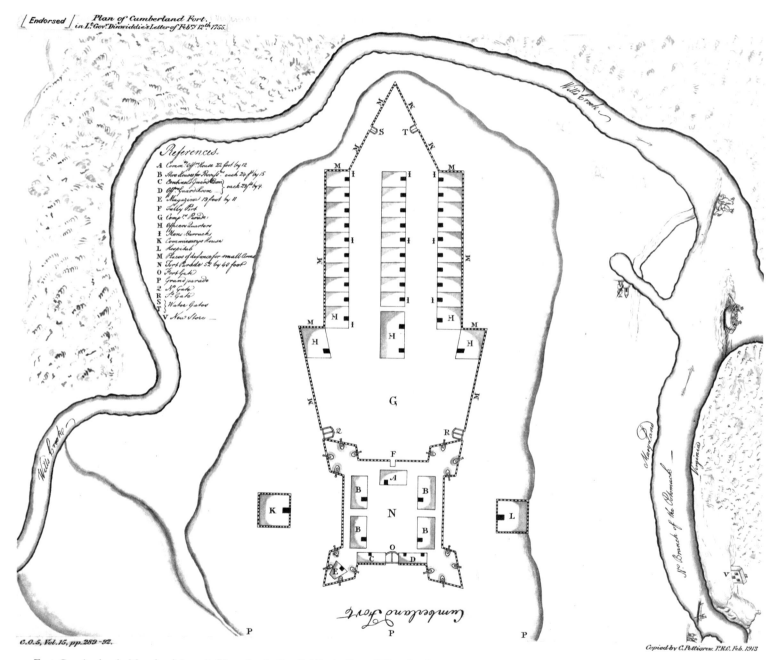

References.

A Comm.d Off.r House 22 feet by 12
B Store House for Provis.t each 24 f.t by 15
C Combined Guard Room } each 24 f.t by 9
D Off.r Guard Room }
E Magazine 13 feet by 11
F Sally Port
G Comp.n Parade
H Officers Quarters
I Mens Barracks
K Commissarys House
L Hospital
M Places of defence for small Arms
N Fort Parade 52 by 40 feet
O Fort Gate
P Grand parade
Q N.o Gate
R S.o Gate
S T } Water Gates
V New Store

Wills Creek

Wills Creek

Maryland

Virginia

S.t Branch of the Potomac

Cumberland Fort

Copied by C. Pettigrew. P.R.C. Feb. 1913

14. Fort Cumberland, Maryland (1754). Plan; facsimile. *Public Archives of Canada, Ottawa.*

15. Fort Cumberland, Maryland (1754). Lithograph (1755). *United States Military Academy, West Point.*

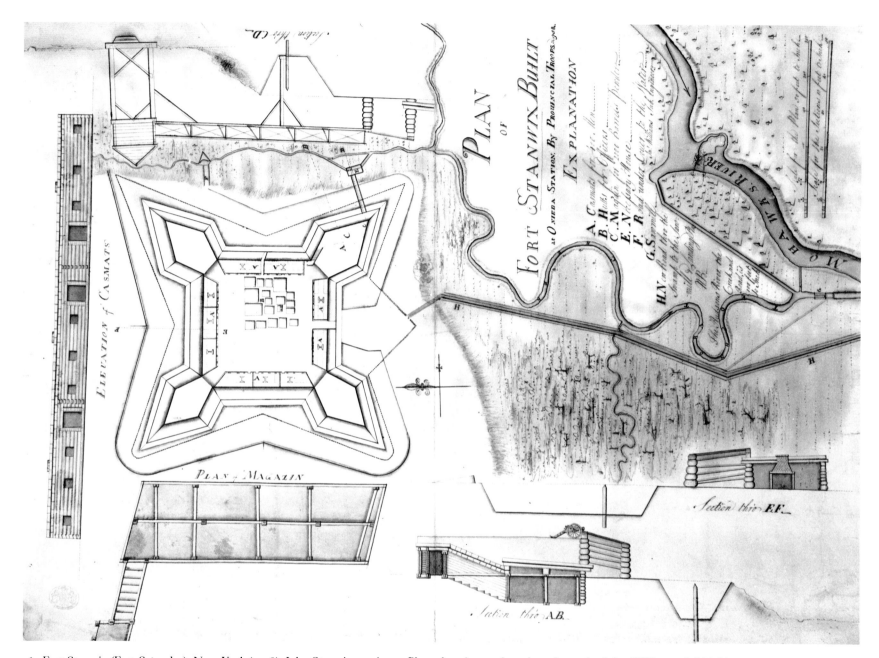

16. Fort Stanwix (Fort Schuyler), New York (1758). John Stanwix, engineer. Plan, elevation, and sections drawn by John Williams. *British Museum, London.*

which was directed by Colonel Innes, was an additional early English example of the square, bastioned plan (figs. 14, 15). Described by George Washington as "not capable of an hour's defence, if the enemy were only to bring a single half-pounder against it,"[42] Cumberland consisted of only a small square enclosure with storehouses and adjunct works around the barracks buildings. Projecting from the main enclosure and forming a part of the fortifications was a hornwork with a series of reentering angles.

Fort William Henry, built in 1755 under the direction of Major William Eyre, also had a rectangular, bastioned trace. Located south of Fort Carillon (Fort Ticonderoga), this strategic defense was a log and earth structure designed to resist a regular siege.[43] The ramparts and parapets were formed from a crib of logs dovetailed together with the spaces between filled with earth. Incorporated into the ramparts were earth-encased magazines and earth-covered enclosures for the security of the men. Minor outworks, consisting of a ditch and a small glacis, supplemented the defenses.

Other strongholds of the 1750s which also conformed to the rectangular, bastioned configuration are the Pennsylvania Forts Halifax (1756) and Augusta (1756), designed by Richard Gridley (1710–96); Fort Ligonier (1758), erected by Brigadier General John Forbes (1710–59); and Fort Stanwix (Fort Schuyler), New York (1758), designed by John Stanwix (fig. 16). Characteristic of forts in the wilderness, the first three were primarily timber works, while the last was of wood and earth.

FIVE-BASTIONED FORTS

The four-bastioned trace and minor variations of it were extensively used in the early development of America because it was a form which was readily defensible by a small garrison and was simple to construct. It was, in theory, limited in size by the length of the curtains and faces of the bastions, since to effectively defend them their combined length had to remain equal to or less than the range of accurate musket fire. However, at mid-century most rectangular or four-bastioned forts measured less than one hundred yards on an exterior side of the polygon, allowing lines of defense that were considerably less than the range of musketry.[44]

Theoretically, when it was necessary to increase the defensive strength of a fort with a larger number of men than could be contained within a square, bastioned fort, a trace based on a polygon with more sides was required. A fort developed from a pentagon or hexagon, with curtains of comparable length, would then enclose more area. However, in America the pentagonal trace was always selected simply because the geometry of the figure adapted to the terrain more readily, not because more interior space was required.

Seven years after the French began work on the first known pentagonal, five-bastioned fort in North America—Fort Beauséjour (Fort Cumberland), New Brunswick (1751–55)—the English used the pentagonal, bastioned form of fortification for Fort Pitt, Pennsylvania (1759). The most extensive work undertaken by the British in America before the Revolution, it was built at the confluence of the Monongahela and Allegheny rivers near the site of Fort Duquesne and was designed by the English engineer Harry Gordon (fig. 17). A pentagonal trace was developed because it conformed to the triangular shape of the site defined by the two rivers;[45] if the engineer had desired, the required area for the parade could easily have been enclosed by a four-bastioned enceinte with satisfactory lines of defense.

Like the French Fort Carillon, which had outworks to defend the lake against ships, Fort Pitt had, overlooking the river, outworks which were effectively designed to adapt the

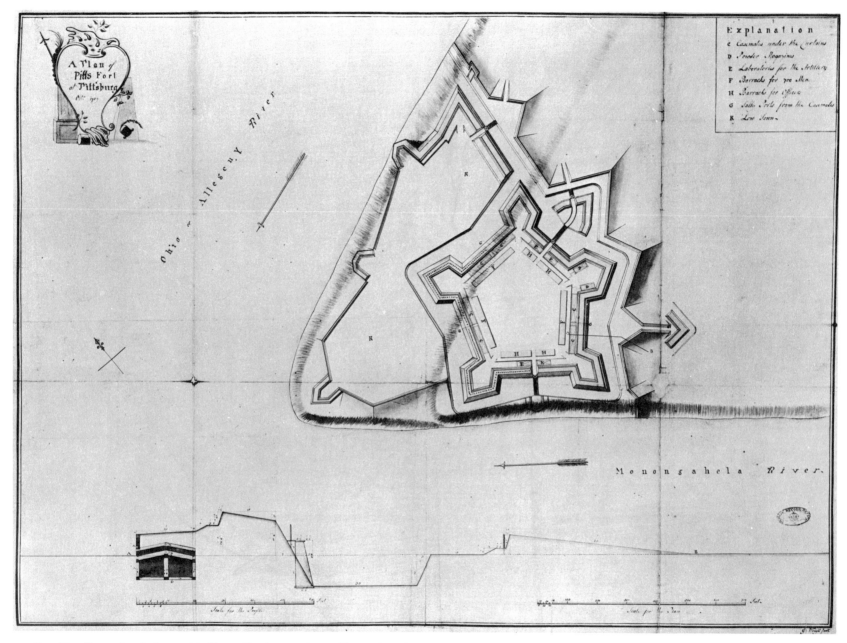

A Plan of
Pitts Fort
at Pittsburg

Explanation

C Casemates under the Curtains
D Powder Magazines
E Laboratories for the Artillery
F Barracks for 700 Men
H Barracks for Officers
G Sallie Ports from the Casemates
K Low Town

Ohio or Allegeny River

Monongahela River

Scale for the Profile

Scale for the Plan

17. Fort Pitt, Pennsylvania (1759–65). Harry Gordon, engineer. Plan and profile (1759) drawn by G. Wright. *Public Records Office, London.*

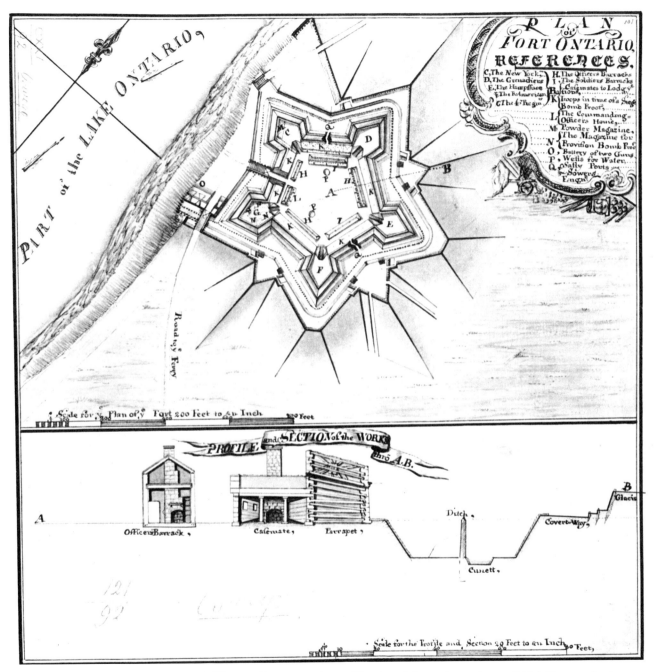

18. Fort Ontario, New York (1759–63). Thomas Sowers, engineer. Plan (1759) drawn by engineer. *British Museum, London.*

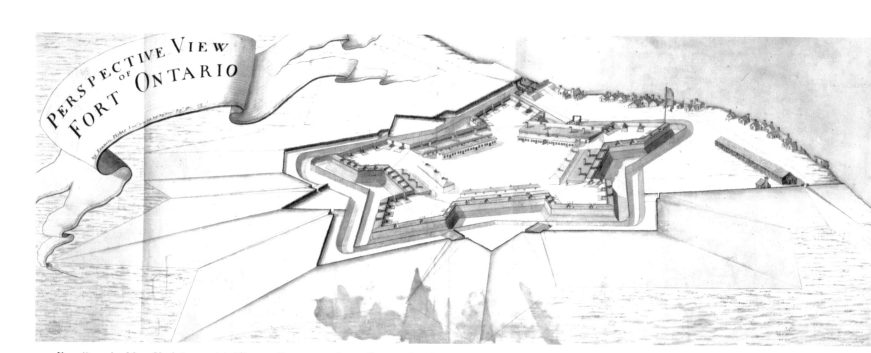

19. Fort Ontario, New York (1759–63). Thomas Sowers, engineer. Perspective view (1761) drawn by Francis Pfister. *British Museum, London.*

geometrical regularity of the pentagonal trace to the irregularity of the perimeter of the site. Developed according to the judgment of the designer, rather than in conformance to formulated designs, these outworks were adjusted to particular conditions of exposure to anticipated attacks as influenced by uneven terrain.

Pitt was built with earth and masonry. The ramparts on the landward side only were strengthened by a fifteen-foot brick scarp capped by a stone cordon. Above and behind the scarp, according to established practices, earth was used to form the terreplein, banquette, and parapet. On the landward side of the triangular site the defenses were augmented by two places of arms, a ravelin, and a ditch.

The five-bastioned, pentagonal design was also selected by

the English for Fort Ontario (1759–63) in response to an analysis of the terrain (fig. 18). Located at the junction of the Oswego River and Lake Ontario in the region of several previous fortifications,[46] the work was slightly smaller than Fort Pitt, and the ramparts comprising the pentagonal enceinte were of logs and earth. Casemates, or bombproofs, of timber and earth were contiguous with the curtains, and, in typical fashion, the various barracks were arranged parallel to the polygon of fortification (fig. 19).

In the English fort at Crown Point, New York (1759), a work built of earth and pine logs, the five-bastioned configuration appeared again. A work that was comparable to Forts Pitt and Ontario in area, the geometry of the trace also appears to have been developed to fit the site, an irregular peninsula formed by a bay and the main body of water of Lake Champlain. One front faced a land approach, and one overlooked the lake, while three conformed to the coastline of the bay.

FORTIFIED TOWNS

Although it was not common to fortify entire communities in America, as was the practice in Europe, bastioned enceintes were developed to enclose several coastal towns that, because of their location, were important from a military point of view. This system of fortification was used by the English to circumvallate Frederica, Georgia (1736), a buffer town. Located on Saint Simon Island in the Frederica River less than one hundred miles north of the fortified Saint Augustine, Florida, the town of Frederica was rectangularly platted and surrounded by a ditch and a rampart. The trace of the enclosure was developed on a regular polygon; according to James Oglethorpe—planner of Savannah, Georgia (1773), which was also fortified—it was designed as "half a Hexagon with two Bastions and two half Bastions and Towers after Monsieur Vauban's method upon the point of each Bastion. The walls are of earth faced with Timber, 10 foot high in the lowest place and in the highest 13 and the Timbers from eight inches to twelve inches thick. There is a wet ditch 10 foot wide."[47] The five-sided towers were two stories high and were designed for the mounting of cannon.

A citadel was located on the bank of the river on an axis with the main gate at the end of a broad street (fig. 20). It was a four-bastioned earthwork which first had sod revetments and which later, in 1739, because of the instability of the sod, was revetted with blue clay.[48] Since the sloping ramparts of earthwork fortifications could have been easily stormed, various palisades and stockades were constructed to prevent surprise attacks.

Even with all these fortifications, Frederica was relatively weak. This evidently did not go unnoticed at the time, for one observer, Harman Verelst, remarked that "The real defense of the town is the Woods and Swamps."[49] However, the defenses were never tested, for in the 1742 invasion the English defeated the Spanish outside the town.

In the expansive occupation of the continent, French colonial activity also included the planning of several fortified towns. Among these outposts were the towns of New Orleans and New Biloxi, both of which were located on the Gulf Coast. New Orleans, Louisiana, was founded in 1718 on a location which had been selected by Le Moyne de Bienville and naturally consisted at first of only a few small, crude structures. Although the original plan design by Sieur de Périer was pentagonal,[50] the town, as finally laid out in 1721 by Sieur de Pauger, was a simple gridiron eleven blocks long and four blocks wide.

Like many settlements in the eighteenth century, New Orleans went through several military crises, and the fortifications varied accordingly. After the 1729 massacre at Fort Rosalie (1716), construction was immediately commenced on an enceinte, a simple stockade with small blockhouses at the corners.[51] However, in the comparatively peaceful period that followed, M. Dumont noted (ca. 1735) that the settlement could justly be called a city and that it wanted only fortifications, which had not yet been begun.[52] Later, in 1760, under the direction of the engineer De Verges, more work of a minor nature was done on the ditches, bastions, platforms, and curtains.[53] Then in 1792, under the Spanish occupation, new plans for the fortification of New Orleans were proposed by Gilberto Guillemard.[54] However, the fortifications were never developed into formidable works.

To the east, elaborate proposals for the town of New Biloxi, Mississippi, were the subjects of several drawings made in

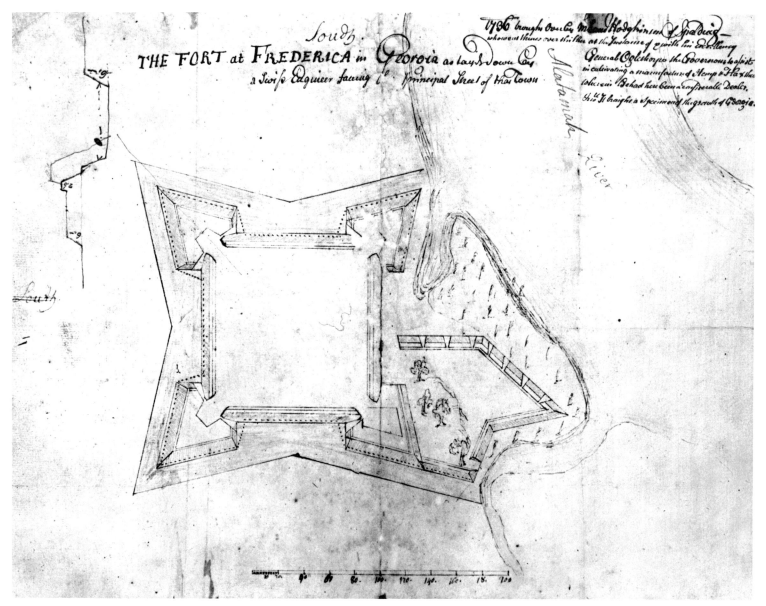

Souдh

THE FORT at FREDERICA in Georgia as layd down by
a Swiss Engineer facing the principal Street of that Town

1736 brought Over by Mr. Hodgkinson of Spalding
who sent these over thither at the Instance of & with his Excellency
General Oglethorpe the Governour to asist
in cultivating a manufacture of Hemp & Flax their
solcin Behad here been a considerable Dealer.
this it maight a Specimen of the growth of Georgia.

Alatamah River

South

20. Citadel, Frederica, Georgia (1736). Samuel Augspourguer, engineer. Plan (1736). *John Carter Brown Library, Providence, R.I.*

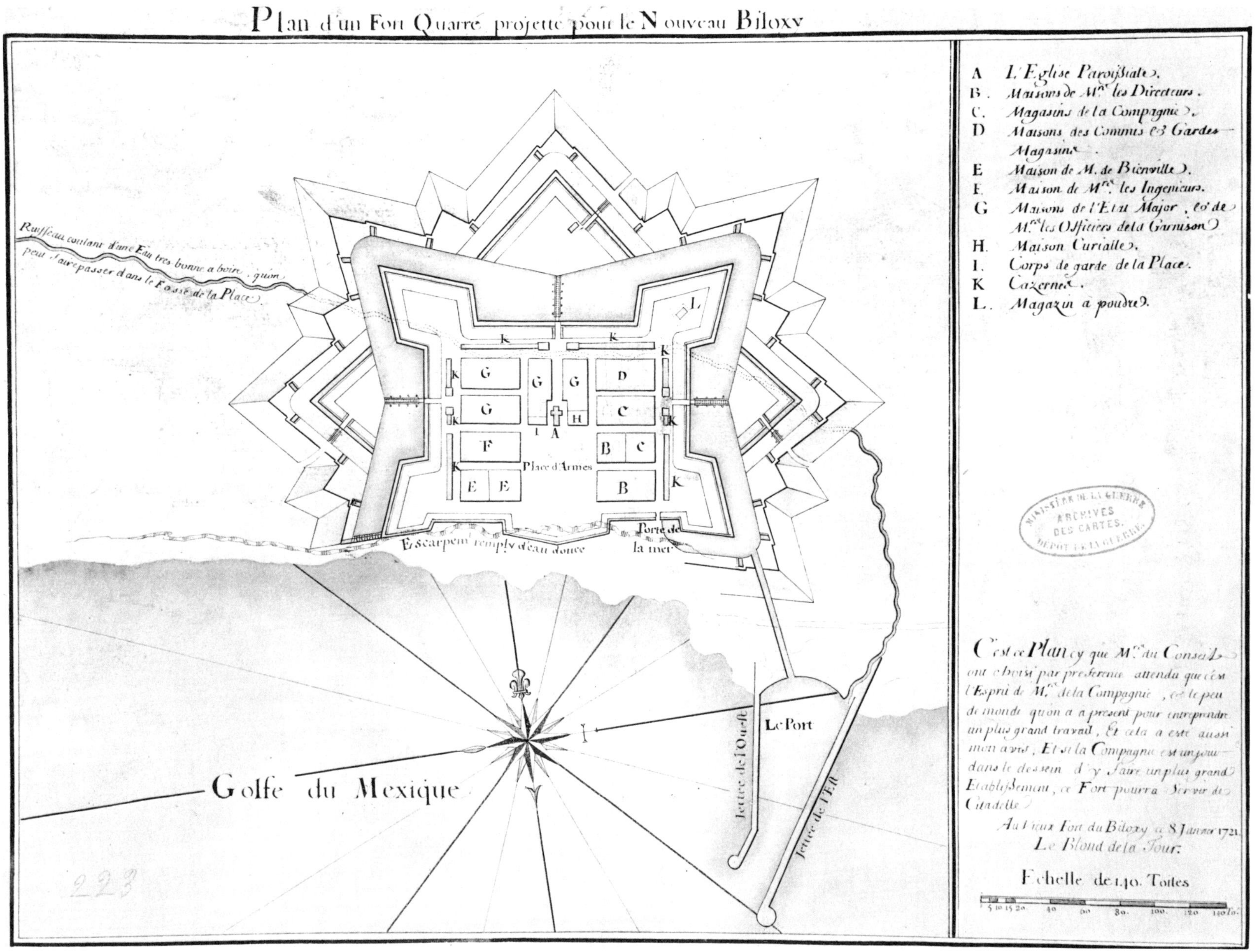

A — L'Eglise Paroissiale.
B — Maisons de M.rs les Directeurs.
C — Magasins de la Compagnie.
D — Maisons des Commis & Gardes Magasins.
E — Maison de M. de Bienville.
F — Maison de M.rs les Ingenieurs.
G — Maisons de l'Etat Major, & de M.rs les Officiers de la Garnison.
H — Maison Curiale.
I — Corps de garde de la Place.
K — Cazerne.
L — Magazin à poudre.

Ruisseau coulant d'une Eau très bonne à boire, qu'on pour faire passer dans le Fossé de la Place.

L
k — k
G — D
G — G — c
H
F — c
I — A — B — c
Place d'Armes
E — E — B
K

Escarpem.t remply d'eau douce

Porte de la mer

MINISTERE DE LA GUERRE
ARCHIVES
DES CARTES
DEPOT DE LA GUERRE

C'est ce Plan cy que M.rs du Conseil ont choisi par preference attendu que c'est l'Esprit de M.rs de la Compagnie, & le peu de monde qu'on a à present pour entreprendre un plus grand travail, Et cela a esté aussi mon avis, Et si la Compagnie est un jour dans le dessein d'y faire un plus grand Etablissement, ce Fort pourra servir de Citadelle.

Au vieux fort du Biloxy le 8 Janvier 1721.
Le Blond de la Tour.

Echelle de 140. Toises.
5 10 15 20   40   60   80   100.   120   140 To:

Golfe du Mexique

Le Port
Jettée de l'Ouest
Jettée de l'Est

223

21. New Biloxi, Mississippi. Le Blond de la Tour, engineer. *Archives du Ministere des Armées, Vincennes.*

1721 by Le Blond de la Tour, engineer-in-chief of Louisiana. The different plans for the new town, which was to replace the Old Biloxi destroyed by fire in 1719, varied in size but were similar in concept to the plan of New Orleans. In all, the *place d'armes*, which functioned as a place of assembly, formed the focal point and was adjacent to the water. The parish church was opposite this open space, was located on an axis, and terminated the vista from the main gate. The town plans were then symmetrically developed around these features.

As the plans for New Biloxi varied in size, they also varied in trace of defensive works, the configuration of the polygon of fortification becoming more complex as the enclosed area was increased, but all embodied the French system of fortification. The rectangular plan and enclosure that were approved—only slightly larger than the smallest of four proposals[55]—like the others had a bastioned trace with ravelins protecting each of the landward curtains (fig. 21). The main difference between the fronts on Le Blond de la Tour's plan and those of Vauban's first system was the omission of tenailles by the former. Although these fortifications were considered important, they were never realized.

Thus, whether in a fortified town or a wilderness fort, the military men who were instrumental in developing works to defend the claims of their respective countries brought across the ocean the bastioned system of fortification that had been developed in Europe. Because of the strength inherent in the regular geometrical relationships of this system, it was widely used by the three major colonial powers. Flexible in size and construction, the four-bastioned fort was frequently employed in diverse locations in America. The first defenses naturally enclosed small areas, but as competition for the continent intensified, fortifications became larger and more formidable.

Although the English, French, and Spanish colonies produced some permanent bastioned works of masonry, most fortifications before the end of the struggle for empire were constructed expediently and lacked durability. Built of wood and earth, they yielded readily to the destructive forces of fire, weather, and war. While they served their times and circumstances, it remained for the next century to witness the development of an America where numerous formidable permanent fortifications could be produced.

2 Transitional Work

Throughout the early decades of the colonization of North America, the English colonies received little help from the mother country in the construction of works for defense. They looked after their own needs, and what they accomplished was largely the result of collective initiative. The forts of these independent colonies in the early years were therefore expedient constructions, since there was neither money nor manpower to spend on permanence.

It was not until the French and Indian War that England belatedly recognized the need to finance the building and garrisoning of permanent fortifications. During that conflict, construction was begun on Forts Pitt, Ontario, and Crown Point, all of which occupied strategic locations. The Treaty of Paris (1736), by which the English acquired the Spanish forts in Florida and all the military works previously established by the French except New Orleans, vastly increased England's inventory of strongholds in North America. To keep Indians and a conquered population in check, many of these newly acquired forts, along with those that had been built by England, had to be garrisoned and maintained in condition for defense.

ENGLISH DEFENSES, 1763–76

The forts acquired from the French were in various states of disrepair; several key works were already in ruin, while others had become dilapidated through neglect. Fort Carillon, for example, had been partially destroyed by the French when they abandoned it, although repairs were made after English troops oocupied it. Wooden revetting was replaced with stone, but then was not maintained. "By the early 1770's the walls had collapsed in places, the parapets which sheltered the cannon on the ramparts had disintegrated into a mass of dirt and rotten wood, while the stone barracks that once held a garrison of four hundred men were... in sad repair."[1]

The forts which the English had built themselves were likewise in poor repair. In 1773 the fort at Crown Point was severely damaged when the wooden revetting, which had been coated with tar to resist decay, burned after a washer-woman's fire got out of control.[2] Fort Pitt was heavily damaged by floods in 1762 and 1763 before its full completion. During the Pontiac War it was necessary to finish it with makeshift barriers of planks, barrels of earth, and bales of deerskin.[3]

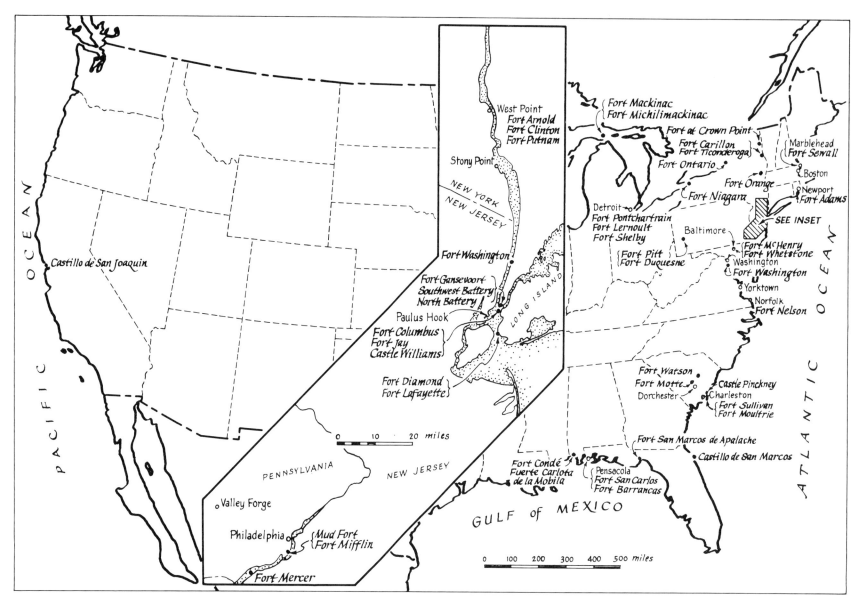

West Point
Fort Arnold
Fort Clinton
Fort Putnam

Stony Point

NEW YORK
NEW JERSEY

Fort Washington

Fort Gansevoort
Southwest Battery
North Battery

Paulus Hook

Fort Columbus
Fort Jay
Castle Williams

LONG ISLAND

Fort Diamond
Fort Lafayette

0   10   20 miles

PENNSYLVANIA

NEW JERSEY

Valley Forge

Philadelphia
Mud Fort
Fort Mifflin

Fort Mercer

PACIFIC OCEAN

Castillo de San Joaquin

Fort Mackinac
Fort Michilimackinac

Fort at Crown Point

Fort Carillon
Fort Ticonderoga

Marblehead
Fort Sewall

Fort Ontario

Boston

Fort Orange

Fort Niagara

Newport
Fort Adams

SEE INSET

Detroit
Fort Pontchartrain
Fort Lernoult
Fort Shelby

Baltimore

Fort Pitt
Fort Duquesne

Fort McHenry
Fort Whetstone
Washington
Fort Washington

Yorktown

Norfolk
Fort Nelson

Fort Watson
Fort Motte

Castle Pinckney

Dorchester

Charleston
Fort Sullivan
Fort Moultrie

Fort San Marcos de Apalache

Castillo de San Marcos

Fort Condé
Fuerte Carlota
de la Mobila

Pensacola
Fort San Carlos
Fort Barrancas

GULF of MEXICO

ATLANTIC OCEAN

0   100   200   300   400   500 miles

Transitional Work

22. Redoubt, Fort Niagara, New York (1770). Watercolor. *Buffalo and Erie County Historical Society, Buffalo, N.Y.*

In 1763 construction was commenced on redoubts in front of three sides of Fort Pitt which had become incapacitated for defense because of flood damage, and the following year two more units were added. In plan, one of these Fort Pitt redoubts—probably typical of all—was five-sided, with three of the sides square and the other two forming a salient angle pointing outward. These works were not designed to defend against or be defended by artillery. Rifle loopholes only, cut in horizontal timbers, were provided on both stories.

Other redoubts were constructed by the English at Fort Niagara. Just over ten years after England won the fort, the earthworks there were strengthened with works of masonry placed near the gorges of each demibastion, inside the enceinte instead of outside as at Fort Pitt. The first, the south redoubt (1770), was placed near the gate and provided a second point

for controlling access (fig. 22). The second, the east redoubt (1771), was similar but in a more isolated position. Both had massive walls, five feet thick at the base, which tapered gracefully inward to a stone stringcourse near the top. Loopholes were provided in the walls of each of the two lower floors, while cannons were situated behind parapets on the third floor. To avoid casualties from flying splinters, the trussed roof that covered the top floor was designed to be rapidly dismounted in case of attack.[4]

AMERICAN FORTIFICATIONS, 1775–76

The maintenance of old fortifications, the construction of new works—in Canada as well as America—and the keeping of garrisons at widely distributed points by the English demanded a great expenditure from the royal treasury. Since Parliament determined that the American colonies should help defray these operations, revenue acts and, subsequently, coercive acts were passed to raise revenue.

When the rebellious colonials entered into open hostilities in 1775, they possessed only a partially trained militia, few military leaders, no trained engineers, and no fortifications other than those designed to resist Indians. Men who directed the construction of the early defenses so needed by the American army had comparatively limited backgrounds and talents. Before the war was over, their inadequacies would have to be compensated for.

The earliest need for technical direction was answered by Colonel Richard Gridley (1710–96), a military engineer who had seen service in the two sieges of Louisbourg and on the Plains of Abraham in Canada. During his short tenure as chief engineer, he planned a series of works near Boston and directed the construction of entrenchments on Breed's Hill, overlooking Boston, before the battle there in 1775.[5] A few months later, Gridley with another engineer fortified Dorchester Heights, after which the British were forced to evacuate Boston.

Such fortifications as Gridley and other officers of the colonies were able to effect in the early war years responded to immediate needs; they were classed as fieldworks and were impermanent but sometimes effective. Fort Sullivan (Fort Moultrie), South Carolina (1776), for example, was rapidly formed by a double wall of soft palmetto logs filled with sand under the direction of Colonel William Moultrie (1730–1805). Yet to the chagrin of British officers it proved sufficiently strong to repel an attack in 1776 with only minor losses.

Other fortifications were undertaken in the early years of the war at other locations. In the summer of 1776 Fort Washington, New York, was located on an eminence on Manhattan Island and was considered to be quite strong. A pentagonal, bastioned work designed for one thousand men,[6] its effectiveness was verified when it was attacked later that year. It fell, but only to a British force that outnumbered the Americans five to one. Located below Philadelphia on the Delaware River, Fort Mercer, New Jersey (1777)—another fieldwork—proved adequate for one British assault in 1777, but it was taken in another several weeks later. It consisted simply of earthworks ten feet high faced with plank and, to discourage attempts at scaling, it was surrounded by a dense abatis,[7] with a deep ditch between the abatis and the ramparts.

Although the Americans realized some success from their defenses, the need for men with abilities in the art of fortification was yet quite apparent. Fort Ticonderoga had been strengthened by the Americans after they captured it in 1775, but it became untenable when the British assumed a position on nearby Mount Defiance, from which they could command the fort with cannon fire. The defenders had considered the mountain inaccessible and had not fortified it. Their oversight prompted a British officer to conclude "that the Americans have no men of military science."[8]

But the lack of trained engineers in the American army was evident before the abandonment of Fort Ticonderoga in the summer of 1777. George Washington in 1775 clearly saw this need when he wrote from Cambridge that there was "a want of engineers to construct proper works and direct the men."[9] Later Washington lamented, "I have but one on whose judgement I should wish to depend in laying out any work of the least consequence."[10]

CONTRIBUTIONS OF FOREIGN ENGINEERS

Recognizing Washington's need for competent military personnel, Congress late in 1776 instructed the colonial commissioners in France—Benjamin Franklin and Silas Deane—to procure the needed men from the French army. With the sanction of the French government, four engineers were officially engaged to assist the American cause: Louis-Lebèque Duportail (1743–1802),[11] de Laumoy (n.d.), Luis de la Radière (d. 1779), and Jean-Baptiste Gouvion (1747–92).[12] The service of these engineers proved to be invaluable. They advised not only on the construction of many defenses but also on the planning and directing of war operations.

Among the works of Duportail were the fortifications at the Valley Forge Encampment (1777–78). A site on which huts were devised for winter quarters was selected overlooking the Schuylkill River (fig. 23). This location, strong by nature because of steep hills, was then strengthened further with entrenchments, redoubts, and obstacles of brush. When completed, the camp was sufficiently strong to deter an open English attack;[13] it could have been taken by the British only with a regular siege—impossible during the winter months because of the frozen ground.

Other works of defense were set up at other strategic locations. In 1777 the British sent one army from Canada to Lake Champlain and the upper Hudson Valley, another into the Mohawk Valley, and a third up the Hudson River. Under a plan designed to separate the colonies, the three armies were to meet in Albany. The attempt failed, but it proved to Americans, particularly to Washington, the importance of maintaining control of the Hudson Valley. The following year a committee selected West Point as the most desirable location for fortifications for that purpose. The works subsequently constructed there were considered to be among the most important in the Revolution.

Thaddeus Kościuszko (1746–1817), a Polish engineer, was placed in charge of executing the works at West Point, replacing de la Radière, who originally worked on them (figs. 24, 25).

The main objective of the fortification was to prevent British ships from passing.[14] To accomplish this, Fort Arnold (Fort Clinton), an irregularly traced earthwork laid out by de la Radière but completed by Kościuszko, was built at the edge of a flat hill above the Hudson. In 1778, the year it was commenced, it was described as being "six-hundred yards around, twenty-one feet base, fourteen feet high, the *talus* two inches to the foot."[15]

Fort Arnold was supported by batteries located on both sides of the river near the level of the water. Among them were Chain Battery (1778), Lanthorn Battery (1778), and Water Battery (1778), all of which were similar in construction, with battered stone walls of dry masonry. The spaces inside the walls were filled, creating level platforms. On these platforms were parapets of either fascines or log cribs and gravel or earth. Cannons, mounted on wooden platforms, fired *en barbette* or through embrasures in the parapets.[16]

On a commanding hilltop high above all the other works loomed Fort Putnam (1778, 1795), the key to the defensive works at West Point. If the rugged mountain slopes had ever been scaled and this work captured, no other fortification at West Point would have been tenable. The lofty, solid rock apogee which this fort surmounted necessitated an unusually irregular trace. In 1802, following some repairs which were made before the turn of the century, Jonathan Williams described the rugged work: "The rude shape of the Rock on which it stands must have superceded other considerations as to the principles of Fortifications. It has eight salient & nearly as many re-entering angles, some of which are acute, and some obtuse, while one part is a curve."[17]

At the time, the defenses constructed at West Point were the most elaborate and permanent undertaken by the Continental Army. During 1779 some twenty-five hundred men labored there.[18] When the works were completed, Duportail

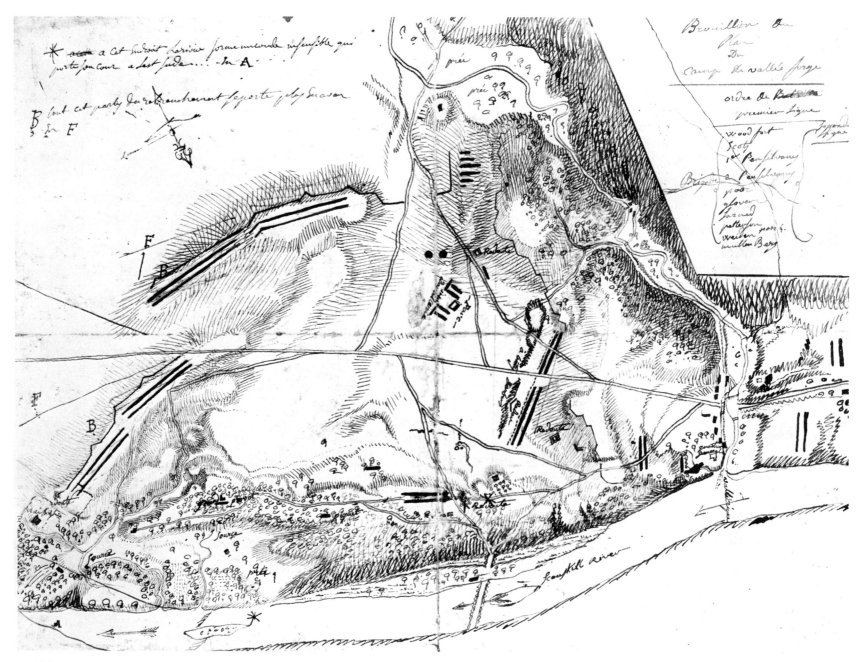

23. Valley Forge Encampment, Pennsylvania (1777). Plan (1778) drawn by Duportail. *Historical Society of Pennsylvania, Philadelphia.*

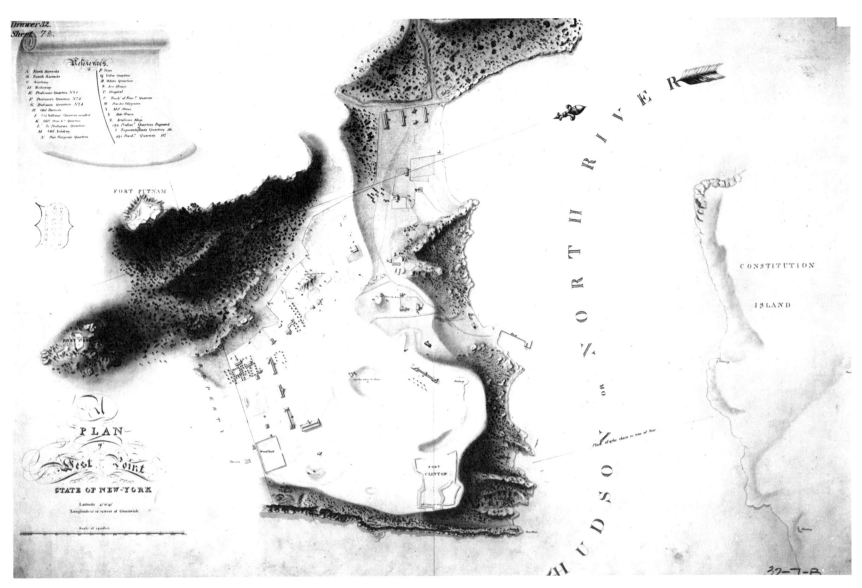

24. West Point, New York (1778). Plan (1818) drawn by George W. Whistler. *National Archives, Washington, D.C.*

25. West Point, New York. Painting (ca. 1870) by Seth Eastman. *Architect of the Capitol, Washington, D.C.*

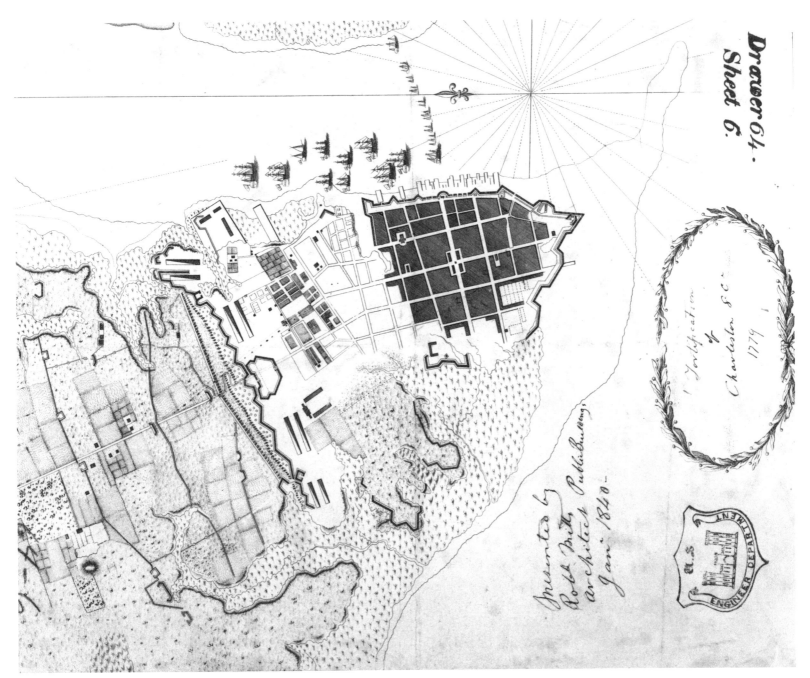

26. Charleston, South Carolina (1779). Plan drawn by Robert Mills. *National Archives, Washington, D.C.*

believed that eighteen hundred troops would have been needed for their effective defense.[19]

To resist the English attack of 1780, French engineers also fortified Charleston for the Americans. Among those seeing service on these works were de Laumoy, Pierre Charles L'Enfant (1754–1825), and Duportail. Working under the pressure of limited time, the French engineers constructed north of Charleston lines of fortifications consisting of pickets, fraises, *trous de loup*, double abatis, and a canal extending from the Ashley to the Cooper rivers (fig. 26). In addition to these works, an attempt was made to close the Cooper River to navigation with a boom "composed of 8 vessels of different size sunk across the channel with Cables and Chains & Sparrs lashed & secured to their lower masts."[20] Although Charleston eventually capitulated, the art of fortification enabled an insufficient garrison to resist for more than six weeks.

HALF-BASTIONED FORTS

Fortifications during the Revolution did not differ in principle from those of previous years. Except in the construction of redoubts, batteries, and fieldworks, where it could not be used, the bastioned trace was employed by all engineers. However, during the war a variation of the regular bastioned system, the half-bastioned trace, appeared in America. In this plan, demibastions replaced conventional flanking devices at the angles of the polygon. Although they had been used on the continent before the Revolution—for example, in a hornwork at Fort Duquesne—there were evidently no forts in which demibastions alone were used.

Known in Europe by the sixteenth century, the half-bastioned trace was inherently weaker defensively than a regular bastioned arrangement. It had fewer flanks, and the faces of the bastions could only be enfiladed from one side. Its main advantage was simplicity, hence, ease of construction.

Although comparatively rare, the half-bastioned plan was used in the 1770s at widely separate locations. The fort at Dorchester, South Carolina (ca. 1775), a small work with lines of defense only slightly over one hundred feet long, had an enceinte which conformed to this configuration. These walls, constructed from tabby, were only three to four feet thick and seven to eight feet high.[21] A much larger and stronger half-bastioned fort was erected at Detroit by British forces in 1778 (fig. 27). Named Fort Lernoult (Fort Shelby) (razed 1826), it was located on a hill overlooking the stockade enclosure of Fort Pontchartrain (Detroit) (1701). An expected attack prompted the construction of the fort, since the earlier stockaded fort could not resist cannon fire. Originally, the engineer, Captain Henry Bird, had planned to construct only a redoubt. Bird reported, "I at first intended only a square (our time as we imagined being but short for fortifying ourselves) but when the square was marked out it appeared to me so naked and insignificant that I added the half Bastions."[22] Bird recognized that a fort with regular bastions was more defensible than a work with fewer flanks,[23] but he knew also that a half-bastioned fort was stronger than a redoubt. This enceinte represented compromise based on the estimated amount of time which was available for construction before an expected attack which never came.

Fort Lernoult was an earthwork with ramparts some eleven feet high and twenty-six feet thick at the base. In typical fashion, the ramparts were perforated with embrasures oriented in various directions to provide uniform coverage of expected enemy approaches. To strengthen the enclosure and to prevent the possibility of the fort being stormed, pickets were planted in the bottom of the ditch, while fraises were

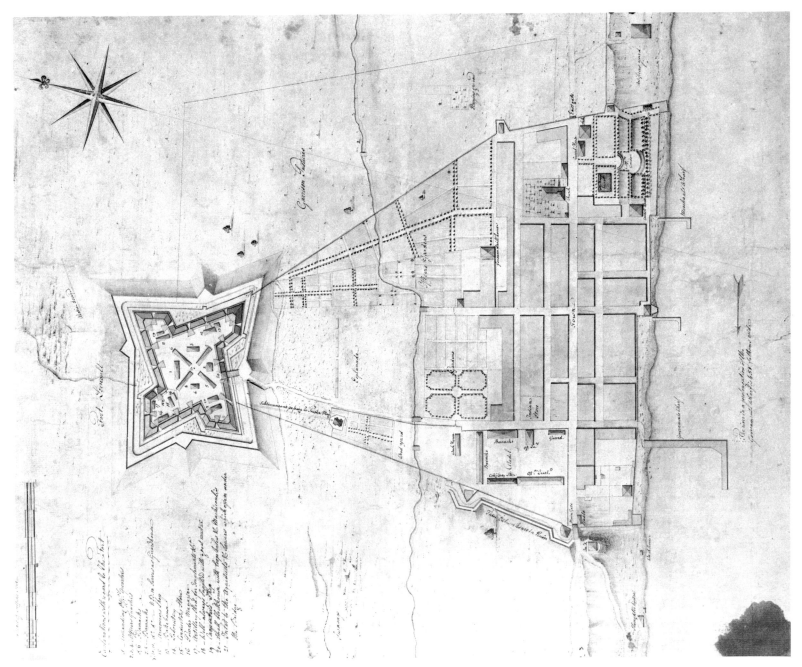

27. Fort Lernoult (Fort Shelby), Michigan (1778). Henry Bird, engineer. Plan (1799) drawn by John Jacob Ulrich Rivardi. *William L. Clements Library, Ann Arbor, Mich.*

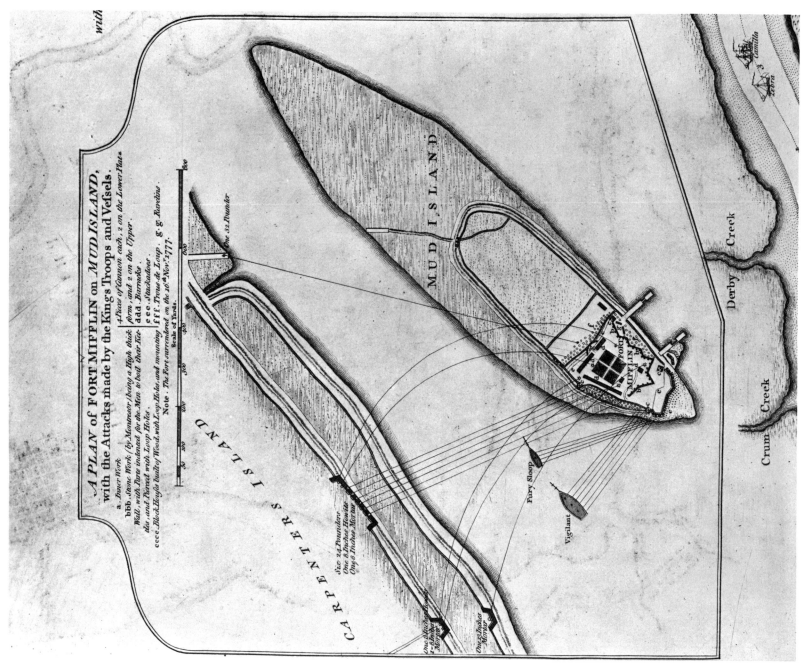

A PLAN of FORT MIFFLIN on MUD ISLAND,
with the Attacks made by the King's Troops and Vessels.

a. Inner Work
b b b . Stone Work (by Montresor) being a High thick
Wall, with Parts indented for the Men to boil their Ket-
tles, and Pierced with Loop Holes.
c c c c Block Houses built of Wood, with Loop Holes, and mounting
4 Pieces of Cannon each, 2 on the Lower Plat=
form, and 2 on the Upper.
d d d . Barracks.
e e e. Stockadoes.
f f f. Trous de Loup. g. g. Ravelin.
Note . The Fort surrendred on the 16.th Nov.r 1777.

Scale of Yards.

One 32 Pounder
One 32 Pounder

MUD ISLAND

FORT MIFFLIN

CARPENTER'S ISLAND

Derby Creek

Crum Creek

3 Camilla

Fury Sloop

Vigilant

Six 24 Pounders
One 8 Inches Howitz
One 8 Inches Mortar

One 13 Inches
& 8 Inches Mortar

One 13 Inches Mortar

28. Mud Fort (Fort Mifflin), Pennsylvania (1772). Captain John Montrésor, engineer. Plan from *Atlas of Battles of the American Revolution,
Together with Maps Shewing the Routes of the British and American Armies . . .* (New York, 1845).

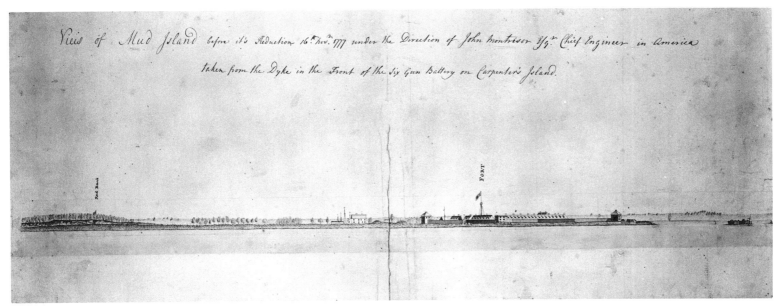

View of Mud Island before it's Reduction 16ᵗʰ Novᵇ 1777 under the Direction of John Montresor Esqʳᵉ Chief Engineer in America taken from the Dyke in the Front of the Six Gun Battery on Carpenter's Island.

29. Mud Fort (Fort Mifflin), Pennsylvania (1772). Captain John Montrésor, engineer. Elevations. *Library of Congress, Washington, D.C.*

placed in the exterior slope of the ramparts. The glacis was terminated at its outer extremity by an abatis.

OTHER BRITISH WORKS DURING AND AFTER THE WAR

Throughout the war years the British continued building or repairing fortifications at strategic locations. Like the American works, these forts responded to a variety of conditions. Some were designed to be permanent, but many others were only temporary, intended to hold a place as its strategic importance developed. The construction on some, started or completed before the war, was completed or altered after the opening of hostilities.

Mud Fort (Fort Mifflin), Pennsylvania, planned in 1771, was started the following year, but it was not completed until 1777 (figs. 28, 29). Located on Mud Island in the hotly contested Delaware River, it was designed by John Montrésor (1736–99),[24] who was appointed chief engineer in America for the British in 1775. The fort consisted of several blockhouses, stockades, and shallow earthworks—not a particularly strong work by design. When they withdrew from Philadelphia, the British evacuated the fort on Mud Island, after which it was occupied by Americans. Shortly thereafter, in six days the British reduced the fort to rubble in one of the most devastating attacks of the Revolution. Ironically, Montrésor had also designed the besiegers' approaches.

The British were also skillful in fortifying with nature. Stony Point—a rocky, 150-foot-high promontory washed on three sides by the Hudson and separated from the mainland by a deep marsh during high tide—was selected as the site for

30. Fort Mackinac, Michigan (1780). Painting (ca. 1870) by Seth Eastman. *Architect of the Capitol, Washington, D.C.*

a stronghold (despite its strengths, however, it was captured in 1779 by soldiers armed with bayonets). Another work for defense strongly fortified by nature was built on Paulus Hook, New Jersey (ca. 1776), a low neck of land separated from the mainland by a deep creek.

In order to strengthen their position in the northern Great Lakes, in 1780 the British under the command of Captain Patrick Sinclair constructed Fort Mackinac, Michigan (fig. 30), on Mackinac Island, another location that was strengthened by the terrain. Designed to replace Fort Michilimackinac, the new structure was an irregular work situated on a high eminence accessible only by means of a steep ramp. The triangular stockaded enclosure was protected with three blockhouses, each with first-story walls of stone and second-story walls of wood.

Toward the end of the Revolution the British attempted to hold southern areas with small, rather weak forts which had been occupied, modified, or newly constructed. Fort Motte, South Carolina (1781), was formerly a residence which was made defensible by the addition of a ditch, earthworks, and abatis; Fort Watson, South Carolina (ca. 1781), built on an old Indian mound, was a stockaded fort reinforced with an abatis; Fort Ninety-Six, South Carolina (ca. 1781), was an earthwork strengthened with a stockade and an abatis. These, along with several others, were all captured by the Americans in 1781.

The British defensive effort in the South finally culminated in October, 1781, behind the fortifications of Yorktown, Virginia (fig. 31). On the heights overlooking the York River, the works consisted of irregularly traced earth ramparts, abatis, fraises, and ditches, all of which were skillfully adapted to the terrain. At the ends of projecting fingers of land, well in front of the ramparts, were placed batteries; still beyond these was a series of redoubts. However, it is well known that these defenses proved inadequate to resist the regular siege of the Americans and Frenchmen. General Charles Cornwallis (1783–1805) capitulated after two parallels had been advanced toward the fortifications.

At the conclusion of the Revolution British efforts at fortification were concentrated in the Great Lakes area. Following the Peace of Paris (1783), England attempted to retain posts at key positions on the American side of the Great Lakes. With a hope that boundaries might change, Forts Lernoult and Niagara were maintained until 1796, and the ramparts of the former were completely rebuilt in the early 1780s. However, under Jay's Treaty these two forts, along with Fort Mackinac, were turned over to the United States.

SPANISH FORTIFICATIONS, 1763–1821

During the period beginning with the end of the French and Indian War and ending with the 1821 cession of Florida, Spain continued the defense of the Southeast. However, no new extensive works were completed. Fort Condé (Fuerte Carlota de la Mobila), built by the French and taken from the English, was maintained but evidently not extensively strengthened. Fort San Marcos de Apalache, Florida (ca. 1759), located on a triangular site at the confluence of the Saint Marks and Warcolly rivers, was under construction for many years but apparently was never finished; when the fort was delivered to the English in 1764 it was not yet half complete. Subsequently, it was abandoned by the British and reoccupied by the Spanish, but it evidently was never capable of withstanding a siege.

It was during this period, after Pensacola, Florida, had changed hands several times and several other forts and blockhouses had been constructed there by the British, that major permanent works, Fort San Carlos and San Antonio Battery (ca. 1787), were begun, the designs of engineer Gilbert Guil-

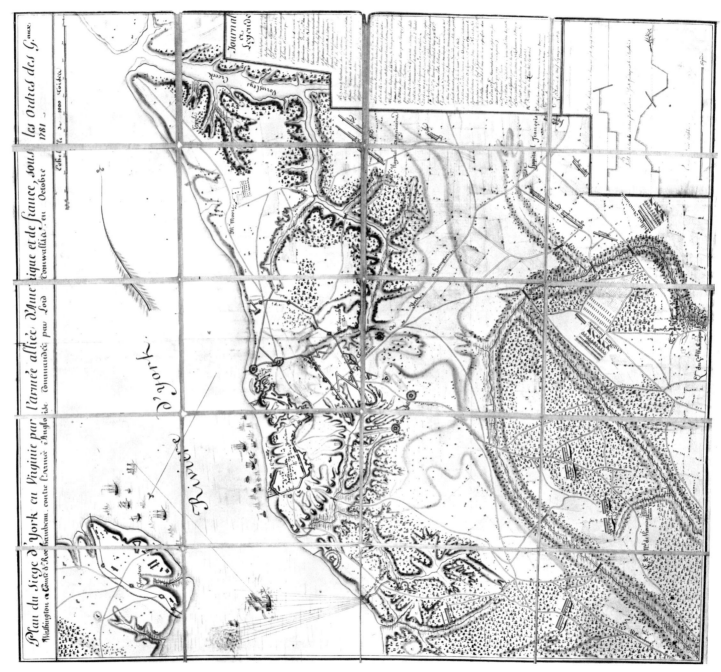

31. Yorktown, Virginia (1781). *John Carter Brown Library, Providence, R.I.*

lemard (fig. 32). However, of the original plan for a four-bastioned fort and a battery, only the latter was completed. To provide wide coverage of the entrance to Pensacola Bay with artillery, it was designed with a "half-moon" trace wherein the semicircular counterscarp and parapet terminated against a high, thick retaining wall in front of which were a magazine and a guardhouse.

The ground that was to have been occupied by the fort was the crest of a high bluff that commanded the battery, now called Fort San Carlos. Since the permanent bastioned work, which was to have been connected to the battery by a tunnel, was never commenced, temporary fortifications were necessary to defend that position against a land approach. An earthwork in the form of a tenaille trace was set up and was connected to the battery by a stockade. Later, this upper work was replaced by the United States with a permanent masonry structure, Fort Barrancas.

Another work with a semicircular trace was built by the Spaniards in California. Begun in 1793 to defend the entrance to San Francisco Bay was Castillo de San Joaquin, a weak work which mounted cannons on a terreplein. Constructed of adobe and revetted brick, it was reported that the enceinte crumbled when cannons were fired,[25] indicating the indefensibility of the work. As did those of England and the United States at the end of the eighteenth century, this fortification reflected the frontier military strength of the government that built it.

American Seacoast Defense: Late Eighteenth Century

Unlike countries of the European continent which were often forced to fortify their inland towns for security against their neighbors, the United States primarily had a long coastline, along with several interior waterways, to protect from other nations. It was realized that any aggressive force would arrive by sea, not by land, and must control a city's harbor before the city could be captured. Hence, the policy of national defense that was developed hinged on fortifying the nation's seaports to restrict the approach of foreign vessels.

In 1794 authorization was given by Congress for the appointment of eight temporary engineers to fortify water approaches to important cities.[26] Among the ports included were those of Baltimore, Charleston, New York, Philadelphia, and Norfolk. The budget for these fortifications, later known as the "first system," was apportioned according to the estimated importance of each harbor. In all cases, however, allotted sums were comparatively small and permitted only the construction of impermanent works. According to specific instructions given to the engineers, fortifications were to consist of batteries, magazines, and either barracks or two-story blockhouses, which were to contain two small cannons on the upper floors. The parapets of the batteries were to be erected "of earth, or where that cannot easily be obtained of an adhesive quality, the parapets may be faced with strong timber, and filled in with . . . earth."[27] To render the batteries more effective, the 1794 orders also included instructions to build a "reverberatory furnace" at each new work for heating cannon balls. Heated on a grate within a large firebox, red-hot cannon balls fired at ships or other targets would ignite them and might even explode them if a lucky hit were made on a powder magazine. Batteries were usually located on both sides of the water so that defenders might annoy with crossfire an enemy attempting to force entry into the harbors. They were placed where the waterways were narrow, yet they were often as far away from the city as possible, since early interception of an enemy was desirable.

The fortifications of the New England seaports financed by the 1794 appropriations also included citadels. Representative

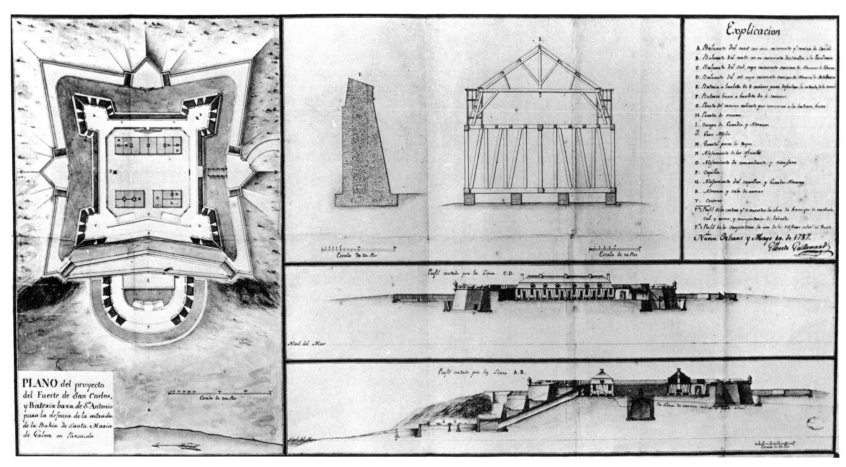

32. Fort San Carlos and San Antonio Battery, Florida (ca. 1787). Plan and sections (1787) drawn by Gilberto Guillemard. *Archivo General de Indias, Seville.*

of these defenses along the North Atlantic coast were the works at Marblehead, Massachusetts, eventually named Fort Sewall and set up under the direction of S. Rochefontaine. According to the engineer's report, these works consisted of a "battery to defend the entrance of the harbor, and citadel, of brick masonry, to defend the battery erected on the spot occupied by the old fort, covering a powder magazine, bomb proof, and calculated for the garrison to live in." The artillery consisted of "six 42, 18, and 24 pounders on coast carriages, and four 6 and 9 pounders, or howitzers, on traveling carriages. Also a reverberatory furnace."[28]

These impermanent defensive works were irregularly garrisoned; some remained incomplete for many years, while others rapidly dilapidated from lack of maintenance. Observing the weaknesses of the coastal works during rising tension with England and France, James McHenry reported, "with respect to foreign nations, it would appear necessary and proper, that the seaboard fortifications should be generally improved, and this defense of our country rendered respectable." He also pointed out the false economy of constructing impermanent works. "A regard to ultimate economy will require, that such of fortifications as may be always important to the general defence should be constructed . . . of the most durable materials."[29] Jonathan Williams later noted that the blockhouses of the first system could easily be beaten to pieces by cannons from enemy ships. At the same time, he further observed that many of the works without bastions were absolutely indefensible.[30]

Charles Pinckney's slogan, "millions for defense but not one cent for tribute," coined, in part, as a response to French demands for bribe money—a paradox in the history of American relations with France—expressed a willingness to increase expenditures for fortifications during 1799 and 1800. With the appropriation of additional money, more defensive works, many of a durable nature, were undertaken. Although the forts were small, the geographical area included was large. When the War of 1812 broke out, there was not a town of any magnitude on the coast that was not protected by at least one battery.[31] The coastal fortifications constructed at the end of the eighteenth century and the beginning of the nineteenth century before 1817 comprised the "second system."

EARLY WORKS OF THE UNITED STATES WITH IRREGULAR TRACES

In keeping with the objectives set forth by McHenry and the correction of defects noted by Williams, important works were begun to defend Philadelphia, Baltimore, Newport, and New York as well as Norfolk and Charleston. Fortifications for Philadelphia were located on Mud Island, the site of defenses constructed by the British in the 1770s. Formidable works for a new fort called, like the one that preceded it, Fort Mifflin were designed in 1794 by Pierre Charles L'Enfant, architect of the city plan for Washington, D.C. (1791). Since L'Enfant's plan for the fort has been lost, it is not certainly known how closely the work that was finally constructed conformed to his design, the first permanent fortification undertaken by the United States (fig. 33). However, it is known that the ambitions of the plan originally far exceeded the budget set by Congress.

Nevertheless, with instructions to complete the fort along the lines started by L'Enfant, Louis de Tousard in 1798 took over the irregularly traced work, which was reported to have been half done two years earlier.[32] Shortly after the turn of the century, Jonathan Williams wrote that the work, which was evidently then completed, consisted of two parts:

33. Fort Mifflin, Pennsylvania (1794). Pierre Charles L'Enfant and others, engineers. Painting (ca. 1870) by Seth Eastman. *Architect of the Capitol, Washington, D.C.*

The first part is formed of four redans defending the river on the south & East. The Walls generally eleven & a quarter feet high are of large stones perfectly cut and put together.... The second part of the works, viz; that which has been aded on the north and West side is composed of two redans toward the West & of a regular front towards the North or land side: that front consists of a casemated Bastion. N.E. a Curtain of ninety six yards & a full half bastion N.W. forming thus a kind of horn work. The wall all round has a good brick revetment, with handsome white Stone Cordon.[33]

Like the work that defended Philadelphia, fortifications for Norfolk, Virginia, were irregularly traced. The works of Fort Nelson (1794), evidently designed by John Jacob Ulrich Rivardi,[34] consisted of an enclosure, traverses, and batteries located behind a parapet that followed the bank of the Elizabeth River (fig. 34). During the stirring times of 1798, Benjamin Henry Latrobe (1764–1820), an English immigrant who directed construction on the national Capitol, was appointed to make a survey of the fortifications of Norfolk. Latrobe then conceived a design for improving the trace of the imperfect Fort Nelson by the addition of bastions, a ravelin, and a wider ditch and by the completion of the traverses shielding the barracks and magazine. Improvements in Fort Nelson along these lines were made in 1802 and several years following.

Latrobe's modifications of Fort Nelson reflect a universal change in the design of ramparts that occurred at the end of the eighteenth century. The original earthworks contained embrasures, through which cannons were fired, framed by wooden joists and faced with planks. Since embrasures often trapped enemy cannonballs and funneled them inward, killing defenders, they were eliminated in favor of a continuous parapet where cannons were mounted on high carriages *en barbette*—a practice adopted by the French for seacoast fortifications during the last part of the eighteenth century.

Unusual traces often resulted not only from the irregular terrain on which the works were situated but also from expediency in incorporating existing works into new construction. Both these limitations influenced Louis Tousard when he designed the first Fort Adams near Newport, Rhode Island. Although he had received instructions to build only a small battery, the importance of Narragansett Roads induced him to propose for the site enclosed works incorporating existing parapets along with barracks, bombproof casemates, a powder magazine, and a furnace.[35] The completed fortification, although hardly formidable, was described as " an irregular star fort of masonry, with an irregular indented work of masonry adjoining it, mounting seventeen heavy guns, a brick magazine, which is too damp for powder. The barracks are of wood and bricks."[36]

Fort Moultrie, South Carolina (1809), a new structure at the entrance to Charleston Harbor to replace the works devised during the Revolution, was a further example of an irregular-trace fort where cannons were mounted *en barbette* (fig. 35). In its completed form the enceinte consisted of a bastion and a demibastion fronting the landward side and an irregular polygon facing the water. Although cannons were mounted to fire through embrasures on the section of the fort defending land approaches, they were mounted to fire over parapets on the sea front.

Developed in response to the analysis of irregular site conditions, the above were but several examples of irregular fortifications. There were other works, some constructed before them and others after, no two of which are alike. However, irregular enceintes were generally more characteristic of impermanent defenses than of durable constructions. In permanent forts, military architects usually favored geometric balance and regularity, at least for the main body of the place.

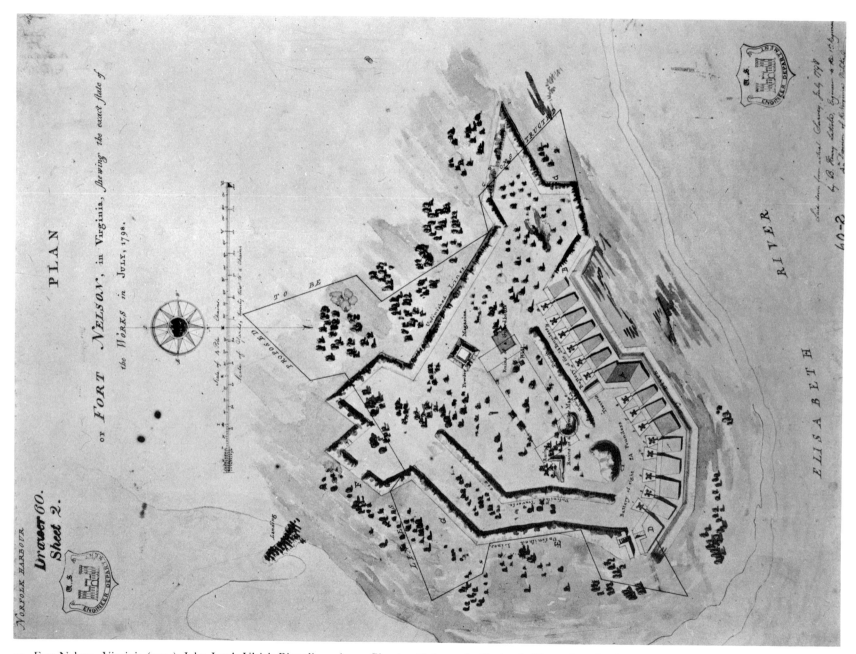

34. Fort Nelson, Virginia (1794). John Jacob Ulrich Rivardi, engineer. Plan (1798) drawn by Benjamin Henry Latrobe. *National Archives, Washington, D.C.*

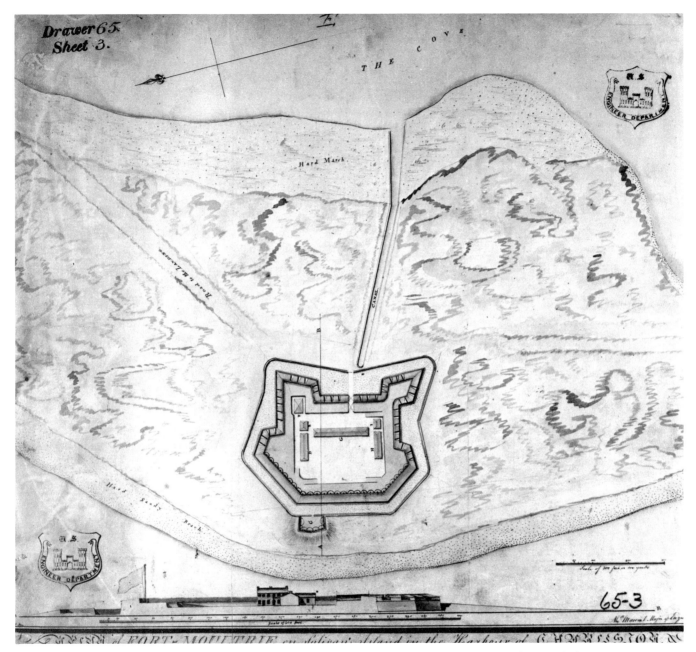

35. Fort Moultrie, South Carolina (1809). Plan drawn by Alexander Macomb. *National Archives, Washington, D.C.*

EARLY FORTS WITH REGULAR TRACES

Regular traces, as it had been observed by Vauban many years earlier, were superior to irregular works. That French military genius also taught, however, that adaptation to the situation was imperative. Likewise, John Michael O'Connor, a nineteenth-century American writer, emphasized that it was the variety of sites that made fortification so difficult a science. Nevertheless, with the use of outworks it was often possible to use regular plans and to adapt to irregularities of the terrain and exigencies of defense.

The geometric regularity so favored by engineers was exemplified by Fort Columbus (Fort Jay), a square, bastioned fortification located on Governor's Island in New York Harbor (fig. 36). The original work, constructed in 1794, was an earthwork with a counterscarp, a gate, and interior buildings, all of masonry. In 1806 the main body of the fort was reconstructed with a stone revetment conforming to a plan[37] evidently by Joseph Mangin. Except for a slight enlargement and the addition of a ravelin with casemated flanks on the north side, the new work followed the same trace as the old.

Mathematical discipline likewise characterized a permanent work for the defense of Baltimore, Fort McHenry. This fortification replaced Fort Whetstone, an earthwork built in 1776 when an attack from the British was anticipated.[38] Whetstone, a star-traced work which was never finished,[39] was considered inherently weak because of its form. When John Jacob Ulrich Rivardi studied the works in 1794, he observed, "that kind of redoubt, [is] always bad in itself, (the fires being oblique, and the salliant, as well as the entrant angles, indefensible)."[40] Needed was a strong bastioned work.

Work on Fort McHenry, a small but durable fortification, was carried on primarily in the years 1798, 1799, and 1800 (figs. 37, 38). As was often the case, many engineers contributed to the design of the defensive works and the buildings within. Succeeding Rivardi before the end of the century were Louis de Tousard,[41] Alexander de Leyritz, and Jean Foncin;[42] the last, a French artillerist and engineer, is usually credited with the design.

Additions and alterations to Fort McHenry occurred over many years. When the fort was immortalized by Francis Scott Key, it was a pentagonal, bastioned fort with a regular trace. The scarp was revetted with brick, while stone was used for the cordon and quoining. To protect the gate, a ravelin was constructed in 1813, but it was replaced in 1837 with the present ravelin, a more formidable work. Later, other outworks were added near the water.

The bastioned Fort McHenry, located between the city of Baltimore and Chesapeake Bay, was designed to defend the Patapsco River. Except in the development of outworks, there was no attempt to modify the architectural form to adapt it specifically to the function of harbor defense. The same bastioned form that had been developed in Europe in response to land-based attacks—and which at Fort McHenry served to defend only a narrow land approach on the northwest—was used to defend against approaching ships on the southeast. In America, new European theory on seacoast defense was yet either unknown or unrespected.

VERTICAL FORTIFICATION

While fortification was a popular subject for many writers during the eighteenth and nineteenth centuries, few specifically covered harbor defense; most discussed fortification in terms of land, not sea, attacks. The problem of defending against ships was much different from that of defending land approaches, since sites for these coastal forts were often on islands or points of land and were restricted in area. Yet the

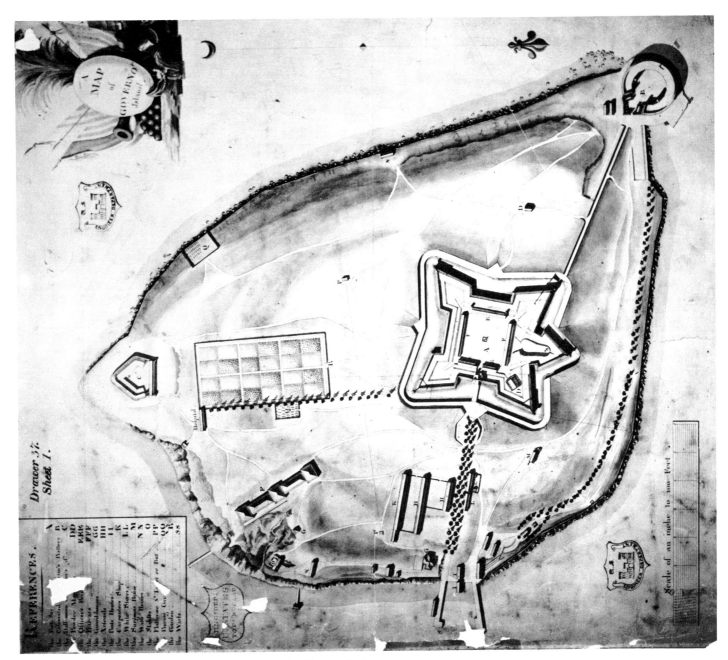

36. Governor's Island, New York. *National Archives, Washington, D.C.*

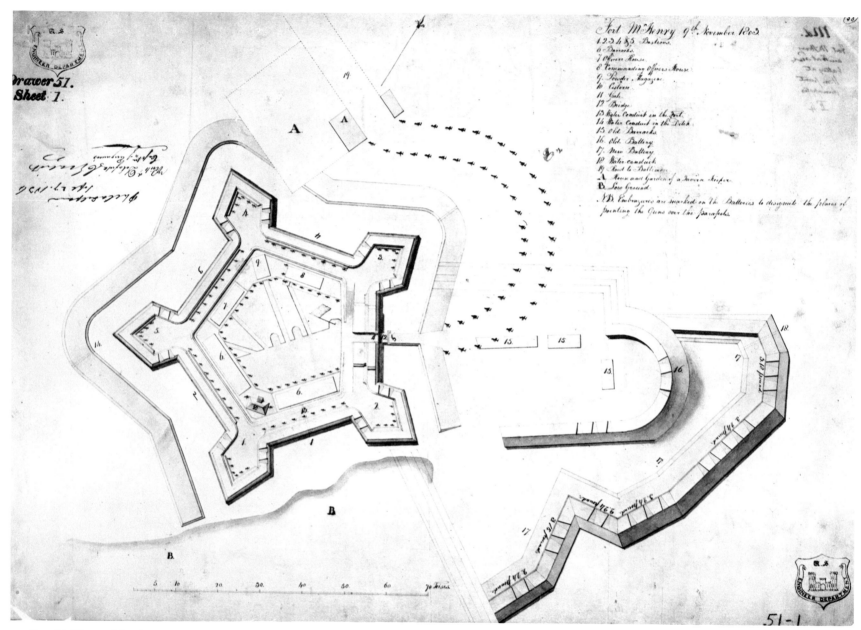

37. Fort McHenry, Maryland (1798). Jean Foncin, engineer. Plan (1803). *National Archives, Washington, D.C.*

FORT McHENRY, BALTIMORE, MD. 1861.

OCCUPIED BY THE 3RD BATTALION OF RIFLES, M.V.M.

Published by J.H.BUFFORD 313 Washington St.Boston

Revenue Cutter
Quarters of Co^s D & B
Capt Dodds Tents
Stables containing
ammunition &c.
Genl Banks' Quarters
Chapel
Light House

9—Sheds containing Ordinance &c
10—Fort McHenry
11—Tents of Staff Officers
12—Fort Carroll
13—Part of Water Battery
14—Co A Worcester City Guards
Worcester Emmett Guards
16—Quarantine

3. Fort McHenry, Maryland (1798). Jean Foncin, engineer. Lithograph by E. S. Lloyd. *Library of Congress, Washington, D.C.*

works that were placed on these locations might be called upon to resist the fire from many heavily armed ships. In order to adequately serve their defensive function, forts had to be strong to resist and armed to retaliate. To encounter ships in open battle, a high concentration of guns was required, necessitating in some cases several levels of guns housed in casemates.[43]

In early American forts, cannons had been mounted only on terrepleins behind embrasured parapets. From the elevated positions of these earth planes, artillerists had a commanding position over the outlying areas, but they were exposed to overhead fire, particularly that of mortars. This danger was eliminated by casemates, the development of which, as essential elements in a system of fortification, is credited to Marquis de Montalembert (1714–1800), a prominent French engineer.[44]

Montalembert combined multiple tiers of casemates within enceintes on circular traces. Such *forts circulaires*, he advocated, were economical, since they required less perimeter of wall for enclosure than did angular traces and required fewer troops to defend them.[45] Moreover, vertical fortifications were easy to adapt to confined sites such as those often encountered along the seacoast.

While they were not the basis of an overall architectural concept, casemates for active defense had been used early in the seventeenth century in colonial forts. Assuming early drawings to be accurate, they were constructed in the flanks of Fort Orange, New York (1617).[46] Several decades later, as previously noted, casemates appeared also in French and Spanish colonial works—Fort Condé, Alabama, and Castillo de San Marcos. But casemates housing cannons were not common in the seventeenth and eighteenth centuries. The protective enclosures, vaulted or trabeated, that appeared were used only for passive defensive functions such as powder magazines or bombproof rooms to which garrisons could retire during bombardment. When located next to the enceinte, the ends of these enclosures were occasionally provided only with small ports for musketry.

Although in theory casemated fortifications provided security for garrisons, cannons, and magazines, in practice there were functional problems with enclosed artillery. Unless there was adequate air circulation, the large quantity of smoke from black gunpowder in the casemate each time a cannon was fired obscured the vision of the artillerists and slowed the serving and aiming of cannons. Therefore, the effectiveness of enclosing cannons hinged largely upon the development of efficient systems of natural ventilation, a matter that occupied military engineers for many years. Among the refinements eventually made in American forts during the early part of the nineteenth century were the development of vents extending up through the casemate vaults and the opening up of the casemates on the parade side of the enceinte. With these improvements air could rush in the embrasure and carry out the smoke through the other openings.

These benefits and problems were observed by an American, Jonathan Williams, first superintendent of the United States Military Academy. Williams served as secretary to Benjamin Franklin in France, where he studied military science and the theories of contemporary French writers on fortification. Following his contact with French works, he introduced into America the circular form of fortification for which the theories of Montalembert appear to have been the inspiration. In the United States this fully casemated, circular type of work was called a "castle."

NINETEENTH-CENTURY CASTLES IN AMERICA

Jonathan Williams designed several castles for harbor defense. Resembling circular shell-keeps of a medieval castle, they were all similar in concept but varied in form according to his interpretation of their specific function. Among these castles were Castle Williams and Fort Gansevoort (Castle Gansevoort) (fig. 39), both a part of the defenses of New York Harbor.

Situated atop a point of rocks at the western extremity of Governor's Island (fig. 36), Castle Williams (1807) was located to create crossfire with Southwest Battery (Castle Clinton) on Manhattan Island and work with Fort Columbus. Named after its designer, it was similar in concept to some French harbor fortifications and had a form which was derived from its function. Castle Williams was a casemated masonry fortification—the first work in America fully casemated for artillery—and was intended to house approximately one hundred pieces of heavy ordnance. Since it had to defend a lateral range in excess of 270 degrees, these cannons were mounted around a nearly complete hollow ring, the discontinuous part of which was filled with adjunct facilities. The containment of this many guns in an exterior diameter of 210

39. Fort Gansevoort (Castle Gansevoort), New York (1808). Lithograph by G. Hayward. *Library of Congress, Washington, D.C.*

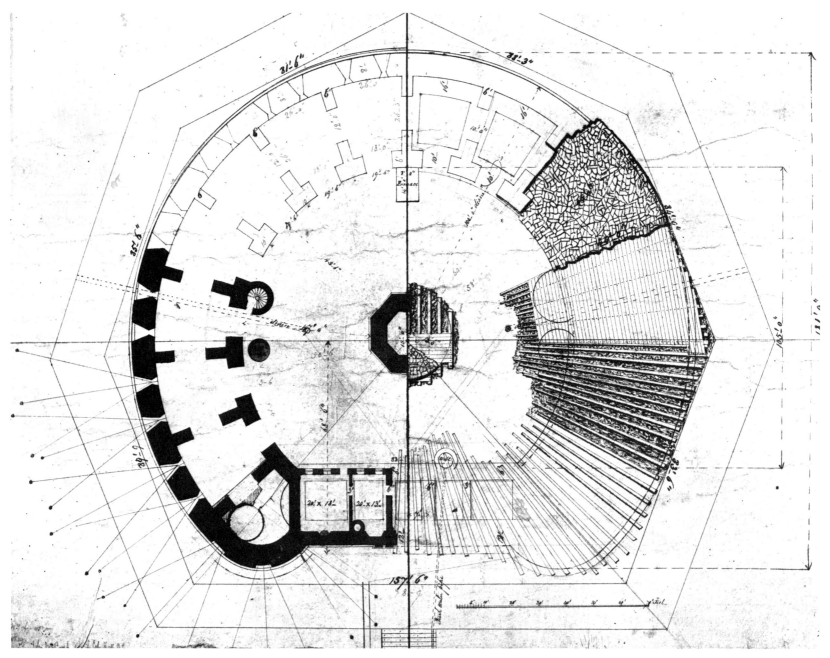

40. Southwest Battery (Castle Clinton), New York (1808). Jonathan Williams, engineer; John McComb, architect. Plan (1808) drawn by architect. *New York Historical Society, New York City.*

feet required three tiers, two of which were vaulted bombproof enclosures.

Southwest Battery (Castle Clinton) (1808), also a part of New York defenses, was located approximately two hundred feet off the shore of Manhattan on an artificial platform in some thirty-five feet of water (fig. 40).[47] Similar to Fort Gansevoort (1808) three miles up the river, its form was beautifully derived from the nature of its function and the physiography of the harbor which it helped to defend. The circular section of the plan provided uniform coverage of the water with cannon fire; at the back where the function changed—embrasures were not needed in the direction of the land—the form changed. The shape of the shoreline was reflected in the graceful manner in which the circle was terminated with a return into the straight segment of the wall.

Another example of this type of fortification was North Battery (ca. 1808) near Hubert Street in New York City (fig. 41). Located in the Hudson River, it was connected to the shore by a drawbridge and contained only sixteen guns.

Charleston Harbor was also fortified with a castle on Shutes Folly Island, the site of a 1798 wood and earth fort which was blown away by the wind in 1804 (fig. 42). Considered at the time of its construction in 1809 to be the most important work in the harbor, Castle Pinckney, South Carolina, was a two-tiered structure built from brick. The lower tier of artillery was contained within casemates, while the upper tier was mounted *en barbette*. As in the castles in New York Harbor, the magazine and other service spaces at Castle Pinckney were located at the back or discontinuous part of the circular form; there were provided a magazine which would " contain two hundred barrels of powder, and quarters for two hundred men and officers."[48]

Other vertical fortifications were built in many forms in conformance to their specific functions. Yet another variation of the castle type of fortification—although not so called—was Fort Diamond (Fort Lafayette), New York (1812), on the Long Island side of the Narrows (figs. 43, 44). The plan was drawn from a square: on the two sides facing the water were two tiers of embrasures; on the remaining two sides were the magazines and quarters.

Other fortifications based on the principles which had determined the forms of the castles were designed after the War of 1812. In 1820 two other fortifications were proposed for the channel at Sandy Hook, New York (fig. 45). They were large, four-tiered masonry structures with rather complex curved traces and were designed to mount nearly two hundred cannons.

All the nineteenth-century castles were located on islands or off shores where there was little opportunity to develop protective outworks. But for that matter there was little need for such outworks, since it was known that they would never be required to resist battering from cannons in a regular land-based siege. Castles thus had only minor flanking structures, if any, and few other provisions for self-defense.

DEVELOPMENT OF CASEMATED FORTS

During the early part of the century, casemates became basic elements of fortification in America not only for castles located on confined sites in the water, but also for forts surrounded by land. Vaulted bombproofs also appeared in Fort Washington, Maryland (1816), the design of multitalented Pierre Charles L'Enfant (fig. 46).[49]

In a 1794 report, L'Enfant set forth his philosophy relative to the importance of strong American fortifications: "All kinds of forts . . . ought . . . to . . . be made capable of self-defence, and should be so situated as to check alone the progress of an enemy. . . . Too much attention cannot be paid, to make all

41. North Battery, New York (ca. 1808). Lithograph by G. Hayward. *Library of Congress, Washington, D.C.*

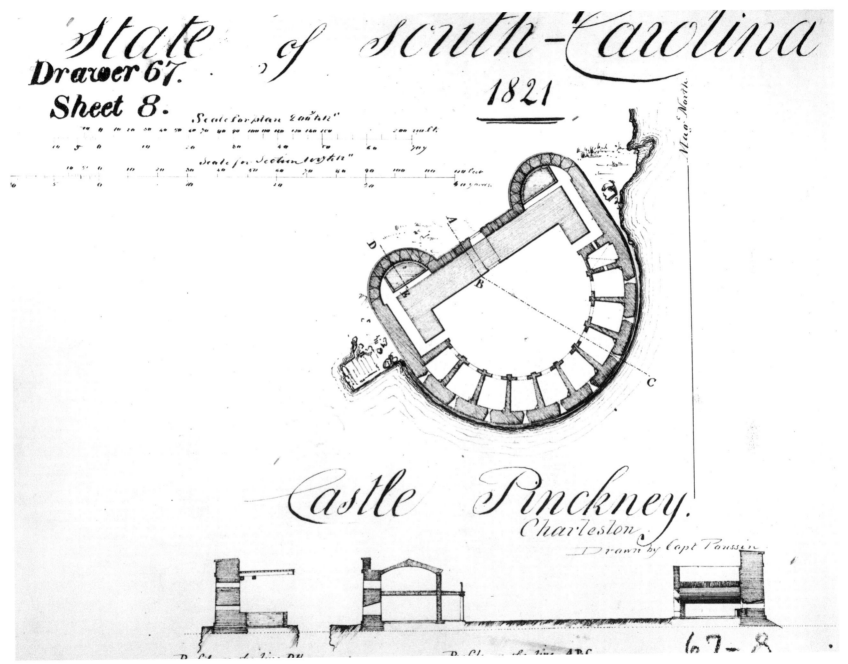

State of South-Carolina

Drawer 67.
Sheet 8.

1821

Mag. North.

Scale for plan 200 feet

Scale for Section 100 feet

A

D

B

C

Castle Pinckney.
Charleston.
Drawn by Capt Poussin

67-8

42. Castle Pinckney, South Carolina (1809). Plan and section (1821) drawn by Guillaume Tell Poussin. *National Archives, Washington, D.C.*

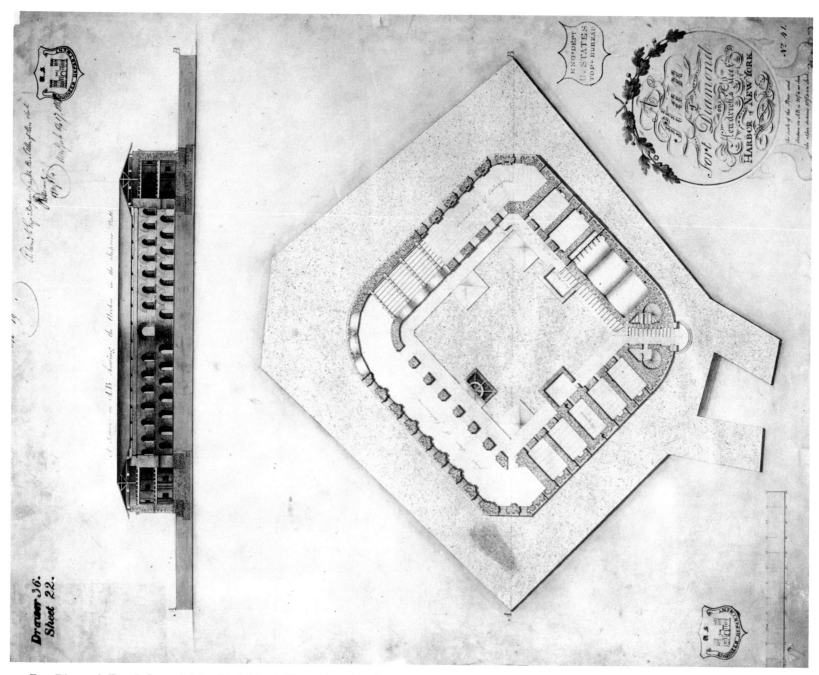

43. Fort Diamond (Fort Lafayette), New York (1812). Plan and section-elevation (1818). *National Archives, Washington, D.C.*

44. Fort Diamond (Fort Lafayette), New York (1812). Painting (ca. 1870) by Seth Eastman. *Architect of the Capitol, Washington, D.C.*

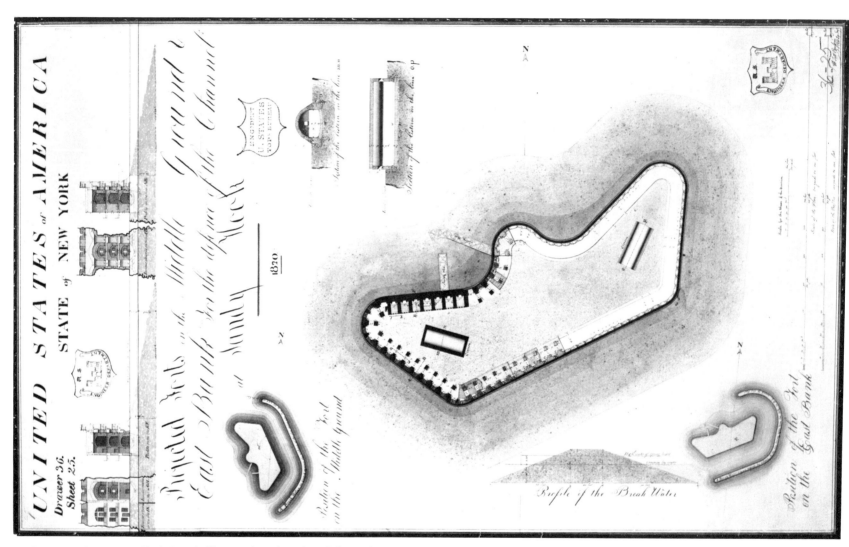

45. Projected forts, New York (1820). Plans and sections (1820) drawn by Guillaume Tell Poussin. *National Archives, Washington, D.C.*

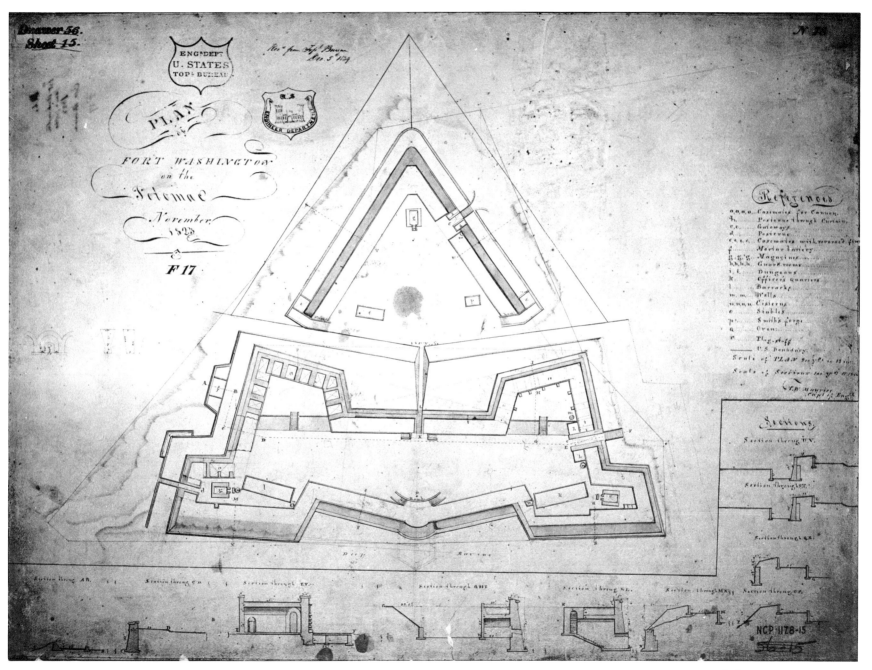

46. Fort Washington, Maryland (1816). Pierre Charles L'Enfant, engineer. Plan (1823) drawn by T. W. Maurice. *National Archives, Washington, D.C.*

fortifications capable of standing against a vigorous attack, and whenever this cannot be done, it is better not to have any, as by becoming useless they must prove greatly prejudicial."[50]

These objectives were fulfilled with Fort Washington, a work designed to defend the national capital. To check the progress of an enemy toward this key city, the fort was located on a high eminence overlooking the main avenue to the city from the sea, the Potomac River. A long curtain was developed parallel to the river to concentrate cannon fire. On this front cannons were mounted *en barbette* to fire on enemy vessels.

To carry out the function of self-defense, a trace was developed to provide flanking coverage for all segments of the enceinte. For enfilading the exceptionally high curtain over-looking the river, two demibastions were provided. Then, as fundamental elements, casemates were used to protect the garrison and the artillery intended for close-in defense from enemy bombardment.

Fort Washington and the other permanent seacoast fortifications of the period 1800–15 terminated an era. Considerable progress had occurred in the art of fortification from the time of British domination through the early part of the century. During these years of transition, although always conforming to scientific theory in the arrangement of components, works of defense by the colonial powers had evolved from comparatively weak constructions of wood into durable works of earth and masonry.

3 The Permanent System

The defensive works of the first system, which were authorized by Congress for the harbors of important coastal cities of the United States, appear to have been constructed where past conflicts had indicated a need; there was no significant national plan based on projected requirements. Those rather deficient and impermanent works were conceived without any regard to defensive interdependence or to their relationship with other important aspects of national defense. In reality, they were nothing more than isolated attempts to provide some form of immediate protection for single points.

In the two decades following 1798 there were significant attempts to improve national defense with the construction of the permanent works of the second system. Yet the program continued to lack depth, strong direction, and coordination. In addition, many fortifications of this period were poorly located; they were too close to the cities they defended. It was all-important that an enemy fleet be intercepted before it reached the immediate vicinity of the city. Because of this, Castle Clinton, for example, was abandoned in 1822 because it was adjacent to New York City. For the protection of Philadelphia, Fort Mifflin became obsolete as a major defense with the construction of Fort Delaware farther down the river.

Likewise, Fort Jackson, Georgia (1808), located on the west bank of the Savannah River and classed as a permanent battery, became a work of secondary importance when other fortifications were built nearer the ocean.

The War of 1812 demonstrated the weaknesses of the fortifications then serving the new nation and had a profound influence on the determination to strengthen the maritime frontier. President James Monroe summarized the impact of the war on the role of fortification:

The policy which induced Congress to decide on and provide for the defence of the coast immediately after the War, was founded on the marked events of that interesting epoch. The vast body of men which it was found necessary to call into the field, through the whole extent of our maritime frontier, and the number who perished by exposure, with the immense expenditure of money and waste of property which followed, were to be traced in an eminent degree to the defenceless condition of the coast. It was to mitigate these evils in future wars, and even for the higher purpose of preventing war itself, that the decision was formed to make the coast, so far as might be practicable, impregnable.[1]

The most feared war was that fought on home ground. With fortifications, it was hoped, war could be excluded from

American territory. Indeed, to prepare for war in tranquil times seemed to be the best means of guarding peace. Therefore, a formidable and permanent system of defenses was imperative.

Motivated by these exigencies, Congress in 1816 authorized the president to employ in the Corps of Engineers a "skillful assistant," or consultant, on fortification. Again, the United States turned to France. To the ambassador who had been assigned the negotiations for this assistant, Lafayette recommended Simon Bernard (1779–1839), former aide-de-camp of Napoleon I and a military engineer of excellent reputation.[2]

The General System

After his appointment and arrival in 1816, Bernard,[3] along with a talented American, Joseph Gilbert Totten (1788–1864),[4] was placed on a board of engineers charged with the creation of a general system of national defense.[5] This board was concerned not only with the defense of individual points but also with logically related groups of fortifications. Consequently, it conceived a great system encompassing the broadest considerations of national geography, military organization, and military architecture. All of these components were considered to be mutually supporting and reciprocally related to each other and to the whole, thereby functioning as a large unit.[6]

A basic determinant of the nature of the general system was the navy. Past conflicts had vividly demonstrated the importance of seapower. Naval depots, harbors of rendezvous, and points of refuge were determined, making up a component system which had to be protected from enemy fleets by fortifications. Moreover, fortifications had the related objective of guarding against blockades by denying an enemy fleet safe anchorage, thus forcing it to ride the sometimes angry sea.

In addition, structures for defense had specific objectives related to military control of the land. The sites selected for forts, by the nature of their positions, were usually strong against attack. By occupying these strong positions themselves, engineers would deprive an enemy of the opportunity to establish bases from which land forces might operate.

The maritime defenses were also planned to strengthen coastal cities against attack. In pursuance of this objective, it was considered desirable to locate them as far away from urban areas as possible in order to force an enemy to land troops at distant points. Then there would be sufficient time to intercept aggressors with a strong force behind field fortifications and to provide an opportunity for various maneuvers such as flanking action or even the separation of the enemy from his fleet.

Intended to function as an integral part of the general system of defense, a system of interior communication was developed in 1824 by Bernard, Totten, and civil engineer John L. Sullivan. With Washington, D.C., as the nucleus, they conceived three interior transportation arteries, by land and water, considered to be important from a military and commercial point of view. One extended from Washington up the Potomac to the Ohio River and thence to Lake Erie; another consisted of a series of canals connecting bays north of Washington; the third was a road that would extend through the Atlantic and Gulf states to New Orleans. These routes were contrived to conduct the necessary supplies to key points for the use of both the forts and the navy, to transfer troops from one place to another, and to preserve domestic commerce during war. The complete realization of this system of transportation, however, was eventually curtailed by the development of the railroads.

Also conceived as a part of the general system of defense was the military organization. There were to be numerous

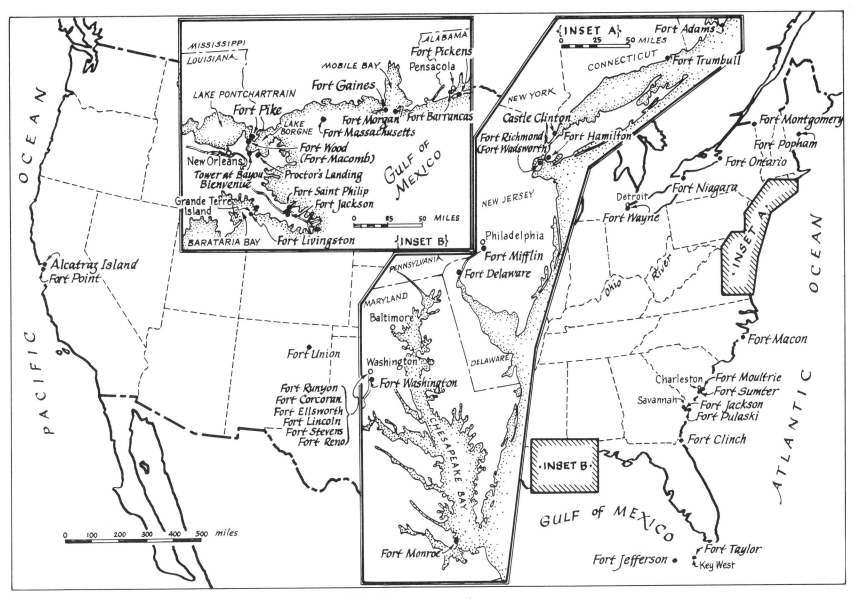

The Permanent System

forts in the completed system, the full-strength garrisoning of which would have been costly. Therefore, the board of engineers thought " every state might organize a certain number of battalions of militia artillery, proportioned to the exigencies and armament of the forts upon its coast."[7] The militia was to live within the call of the forts, and, if an enemy attacked, they were to repair to the fort, where they would remain until danger had passed. This arrangement affected the size of the coastal forts and placed different demands on the garrisons of forts located in various parts of the country, since the potential for a militia in the South was certainly much less than that of the more densely populated East.

No one of these elements in the general system of defense was believed to be much more important than any other; the total constituted the excellence of the system. In reference to the interdependence of navy, fortifications, and transportation, the board of engineers reported, "withdraw the navy and defense becomes merely passive; withdraw interior communications from the system and the navy must cease, in a measure, to be active, for want of supplies, and the fortifications can offer but a feeble resistance for want of timely reinforcements; withdraw fortifications, and there remains only a scattered and naked navy."[8]

The forts that were conceived as a part of the general plan formed what was called the "third" or "permanent" system of national defense. According to their relative importance, they were placed in one of three classes. Forts of the first class were designed to protect the most important commercial cities and harbors, to defend naval arsenals, and to prevent an enemy from establishing a position in this country; with several exceptions, erection of these forts was begun before those of the following two groups. In the second class were forts for protecting cities of secondary rank, some of which were already defended. Forts of the third class, lowest in priority, were

works which would complete the defensive system but which could be deferred without great danger to the nation until the more important defenses were realized. The appropriation for each fort in each class was apportioned according to its projected importance. The date on which construction of a fort was commenced depended, to a great extent, upon the financial condition of the country and the political influence of the section which it would protect.

Since the forts in the comprehensive system of national defense were to be durable and permanent, they were naturally extremely expensive. Consequently, the entire group could be completely realized only over a period of many years. Nonetheless, the board of engineers reported that such fortifications would endure for ages and that the future freedom of the United States might eventually depend upon them. "However long it may be before sensible effects are produced, the result will be certain; and should no danger threaten the republic in our own days, future generations may owe the preservation of their country to the precaution of the forefathers. France was at least fifty years in completing her maritime and interior defences; but France on more than one occasion since the reign of Louis XIV., has been saved by the fortifications erected by his power, and by the genius of Vauban."[9]

These, then, were the general objectives which gave the permanent system of defense its configuration. The interpretation of the specific objectives of each link in the defensive chain gave to each fortification a particular architectural form. The board of engineers, as a body, envisioned the total system, but most of the forts were, in reality, individually planned although several engineers (ex officio members of the board) often contributed to their development.

The plan for each of the forts in the permanent system was generally developed from a specific analysis of two broad defensive functions. Each work was designed to defend a waterway or a position, and, with several exceptions, each was designed to defend against a land-based attack. Thus, although many different forms were developed architecturally, the permanent forts fulfilled several requirements. Every work was designed to contain artillery, garrisons, and magazines which, in the judgment of engineers, would be sufficient to repel an enemy fleet of warships. From the configuration of their plans and profiles the forts were sufficiently strong to resist an open assault, and those which could be approached by land were designed to resist a regular siege for a specified period. Further increasing resistance to attack, they were provided with vaulted bombproofs all covered with masses of earth to protect troops, magazines, and part of their armament, although a large part of the major artillery was mounted *en barbette*. These provisions were clearly expressed in the form of the fortifications designed by Simon Bernard for the Gulf Coast.

WORKS CONFORMING TO THE FRENCH SCHOOL NEAR NEW ORLEANS

In terms of the development of the general system, as well as the designs of individual forts, the creative engineering of Bernard was of great moment to the United States. One of the most talented military engineers ever to see service in North America, he brought from France well-founded knowledge of the French school or system of fortification, which had been derived from the work of Vauban and his successors. He had studied at the École Polytechnique in Paris and had obtained nearly twenty years of experience as a military engineer in Europe. To him was entrusted a key role in the fortification of a country he did not know but whose principles and potential greatness he must have admired.

Following his appointment to the board of engineers, Bernard began his reconnaissance for permanent fortifications in 1817. Working closely with him in the capacity of cartographer and draftsman was Guillaume Tell Poussin (1794–1876),[10] who had been officially appointed to the Engineer Corps as a topographical engineer. In their first year they reconnoitered and prepared maps of the regions around New Orleans and Mobile.

As Bernard explored the terrain, he not only observed locations which advantageously commanded the passes but also visualized the approaches by which an enemy might attack the projected fort. Insofar as commanding positions would allow without jeopardizing their strength, Bernard, along with the other engineers on the board, gave preference to sites where the nature of the ground would hamper the opening of parallels by which an enemy could advance toward the work protected by his trenches. Therefore, when available, sites surrounded by marshes or rocks were selected. It was believed that these sites would be difficult to attack by land, hence monetary savings—as well as security—would be realized through the reduction of the sizes of defensive works. This consideration of terrain was a factor in the design of the forts near New Orleans.

Mostly first-class works because they defended a city of primary importance, the defenses designed by Bernard for the protection of New Orleans were developed as a subsystem within the master plan. Fort Jackson was planned for the south side of Plaquemines Bend of the Mississippi to cooperate with Fort Saint Philip, Louisiana—a fort which had been built ca. 1786 by the Spaniards and improved in 1814 by Americans and which, at the time of the survey of Bernard and Poussin, consisted of a masonry enceinte on an irregular trace (fig. 47). Forts were also projected for Rigolets Pass near Lake Pontchartrain, for Chef Menteur Pass, connecting Lakes

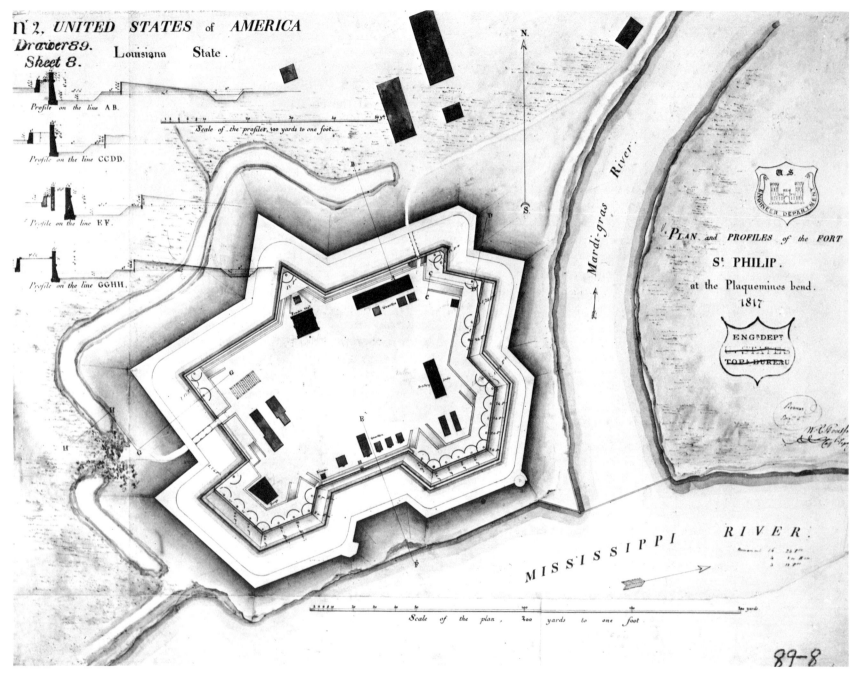

47. Fort Saint Philip, Louisiana (ca. 1786). Plan and profiles (1817) drawn by Guillaume Tell Poussin. *National Archives, Washington, D.C.*

Pontchartrain and Borgne, and for the southern point of Grande Terre Island, overlooking Barataria Pass (fig. 48). At Bayou Bienvenue, a less important point, a three-story tower to mount twenty-four guns was commenced in 1826. All these forts were considered urgent in 1817, all were designed by the French master of fortification, Simon Bernard, and all were consistent in style with the French school.

Fort Pike at the Rigolets Pass (1819), Fort Wood (Fort Macomb) at Chef Menteur (1822), and Fort Livingston at Grande Terre (1841) all had similar functions and were therefore projected by Bernard with virtually identical plans (fig. 49)—although Bernard's plan for Livingston was later discarded. The original form of these forts perfectly reveals the functional considerations which governed their design. They were technically described as "crescent batteries, with the gorges closed by two short fronts of fortification." The round front was derived to provide uniform lateral coverage of the water; since it was believed that ships would not fare well in a duel with a masonry fort, the curved front had no protective earthworks—it was allowed to be exposed in the manner of the castles in the Northeast.

By the time Forts Pike and Wood were designed, history had shown that competently conducted sieges supported with sufficient supplies, adequate artillery, and overwhelming manpower would eventually succeed; unless assistance arrived, it was primarily a question of time until the besieged capitulated. These forts, like all those of the third system, were designed to hold out against a regular siege for from ten to fifty days.[11] This amount of time, it was thought, would allow for the arrival of succor for the defenders.

Successful resistance necessitated outworks with a strong profile: landward, from where besiegers could batter the works with artillery secured in saps, glacis shielded the enceinte; covered ways and salient and reentering places of arms

provided for an outer line of defense; wet ditches and high scarps secured the main body of the places from *coups de main*. Behind the outworks, one regular bastion and two demibastions flanked the curtains. These were the first permanent forts in America wherein the theories of land and water defense were synthesized.

Forts Pike and Wood were casemated works. Thirteen cannons in casemates faced the water passes, while two cannons in each curtain and one in each flank were mounted in casemates to defend the ditch. Built of brick, the bombproofs were roofed with thick barrel vaults, each of which was finished on the top side with inclined planes sloping downward from the crown to shed water into a valley that conducted it to the parade. Typical of American casemated forts, the vaults were then covered with the earth of the parapet and the terreplein, on which was mounted the upper tier of cannons *en barbette*. Each cell was approached from the parade through a long tunnel.

Also conforming to the French school of military architecture was Fort Jackson, Louisiana (1822), another work of Simon Bernard (fig. 50). Although it had a perimeter over twice the length of that of the forts in the vicinity of Lake Pontchartrain, it was considered to be a small work, a condition its designer believed was satisfactory because of the difficulty of attacking it over swampy ground. Jackson was a pentagonal, bastioned fortification with casemates. The regularity of the enceinte was adjusted to the irregular exigencies of defense by the placement of casemates, armaments, and outworks. Curtains facing the water and defending against the passage of ships were casemated for cannons, and, for defense of the enceinte, each flank was also provided with a casemate. The main body of the fort was surrounded on the three landward sides by places of arms and a covered way, the latter of which was provided with tranverses to bar enfilade fire. Two

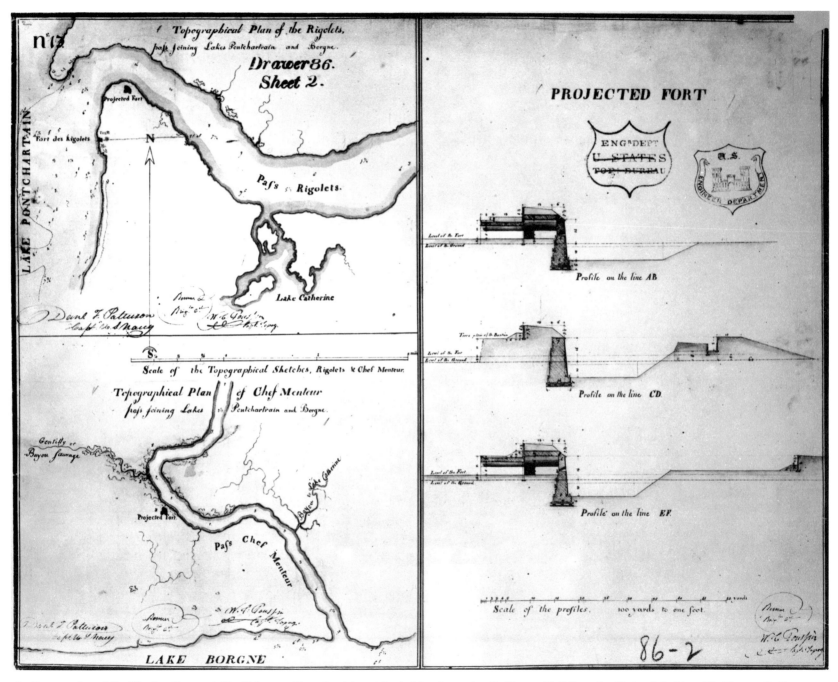

48. Forts projected for Rigolets Pass and Chef Menteur Pass, Louisiana (1817). Map drawn by Guillaume Tell Poussin. *National Archives, Washington, D.C.*

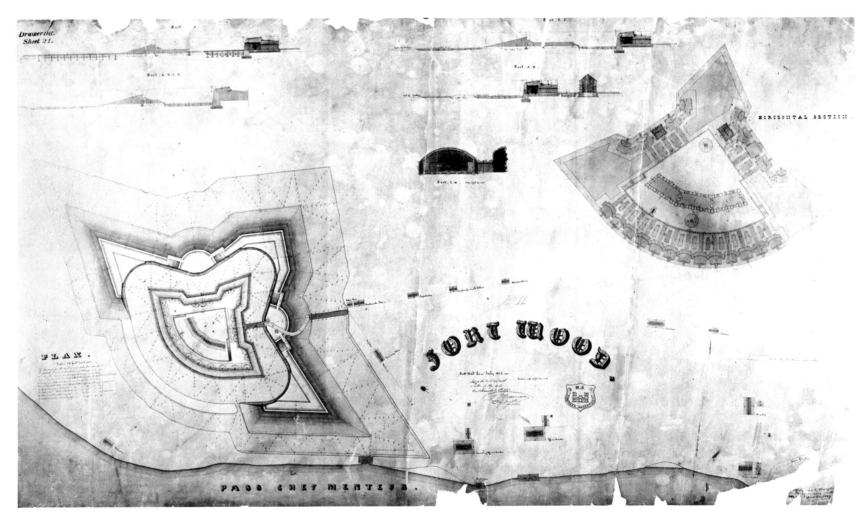

49. Fort Wood (Fort Macomb), Louisiana (1822). Simon Bernard, engineer. Plan, horizontal section, and profiles. *National Archives, Washington, D.C.*

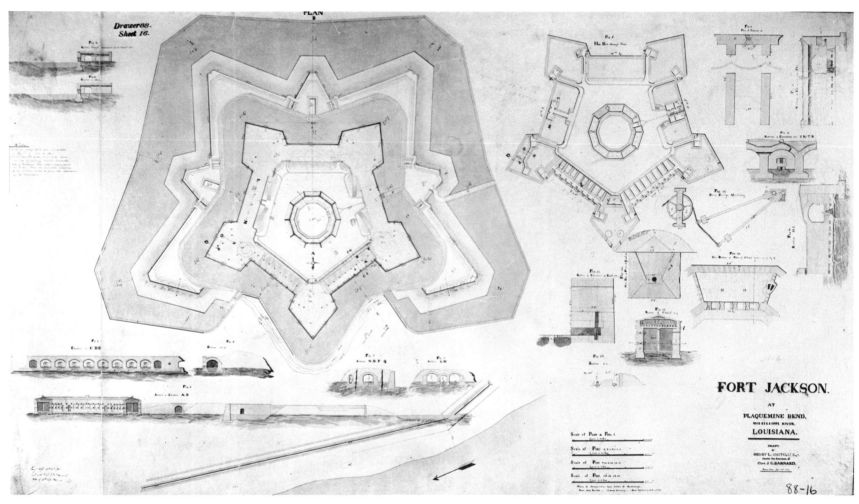

50. Fort Jackson, Louisiana (1822). Simon Bernard, engineer. Plan, sections, and details drawn by Henry L. Smith. *National Archives, Washington, D.C.*

water barriers were included: one between the scarp and counterscarp and the other beyond the rather steeply sloped glacis.

The forts designed for the defense of New Orleans were exemplary of the contemporary French theory on fortification which Bernard incorporated into his work, although the forms of some of the elements were changed by American engineers during construction. For example, the curved section of the counterscarp was established by swinging an arc from the salient of the bastion, and the straight section was placed on a line extending from the point of tangency to the shoulder angle at the crest of the parapet in front of the terreplein: the former was consistent with Vauban's theory while the latter conformed to the work of Louis de Cormontaigne (1692–1752), a disciple of Vauban. Then the parapet walls of the covered way were formed *en crémaillère* around the traverses (in contrast to a series of right-angle indentations as in Vauban's system), another innovation of Cormontaigne (this was a particularly noteworthy feature in some of the larger forts in other sections of the country). In addition, the reentering places of the arms of the forts near New Orleans, as well as those projected for the defense of other cities, were on a circular trace rather than in the form of a salient, reflecting a development made in Europe by Guillaume Henri Dufour.[12] However, in most, if not in all, cases these last features were changed by the superintending engineers back to the traditional form.

Other details of design which were changed were the casemates. For the forts along the Gulf Coast, Bernard designed vaulted enclosures that were short and approached through long narrow tunnels. Perhaps to improve ventilation of gunsmoke, the casemates in most American forts were extended to the parade face and completely opened at the back. Only in Fort Pike and Fort Wood was Bernard's design of casemates evidently retained, thus making them somewhat different from other coastal works.

The forts designed for the defense of New Orleans were all of masonry. Red brick was the basic material, with cut stone used structurally for the cordon, pintle anchors below the embrasures, and traverse circles for cannons and aesthetically to trim the sallyport at Jackson. To support the thick, heavy walls on soft soil, foundations consisted of grillages made from several layers of heavy timbers placed below the waterline—a sound practice, since wood will last indefinitely if continuously kept either wet or dry and not allowed cycles of change.

Works Conforming to the French School around Mobile and Pensacola Bays

After completing their reconnaissance around New Orleans at the end of the second decade, Bernard and Poussin studied Mobile Bay, Alabama, and its role in defense. They concluded that a pair of forts was needed—one at Mobile Point, the other on Dauphin Island—to prevent an enemy from occupying these points and to prevent a blockade (fig. 51). In the scope of the general system, these were second-class forts, considered necessary to secure communication with New Orleans.

The companion forts at the mouth of the bay were conceived by Bernard and both were commenced in 1819. Identical in design, they were pentagonal, bastioned works and were further examples of the French school of fortification. Bernard, speaking through the board of engineers, related his analysis of the forms of the defensive works: "As to the number of sides, the localities required a front to fire upon vessels which approach, one against those passing, and a third against

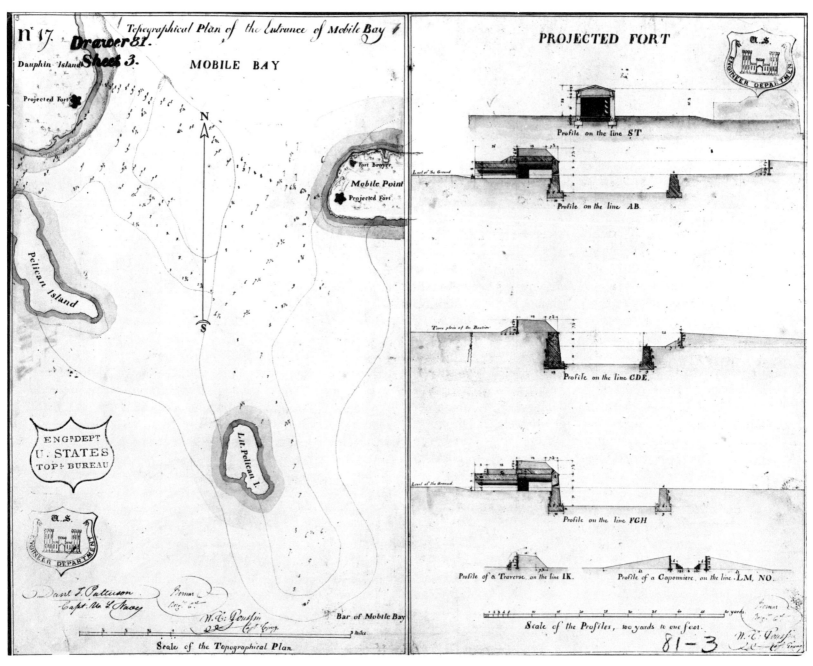

51. Forts projected for Mobile Point and Dauphin Island, Alabama (1819). Simon Bernard, engineer. Map and sections of forts (ca. 1819) drawn by Guillaume Tell Poussin. *National Archives, Washington, D.C.*

such as may have passed; besides, it is indispensible that the fire of two collateral sides should cross upon their capital, as there would otherwise be a space there without defence. These considerations, added to the necessity of providing ample room within for quarters and stores, have produced the pentagonal form."[13]

However, because of a controversy in Congress, work on the fort on Dauphin Island which had been contracted in 1818 was discontinued in 1821 by the government. Although it was considered to be of great importance, construction was not resumed for many years. Then Bernard's plan was scrapped.

Construction on Fort Morgan at Mobile Point proceeded slowly over the years, and it was finally nearly completed at the end of 1833 although work continued thereafter for over a decade. In the trace of the main body of the place it was similar to Fort Jackson and was only slightly larger in size. However, since it was considered to be more vulnerable in a regular siege, and since it was far removed from assistance in case of attack, its dimensions in profile were much greater and thereby stronger than the Plaquemines fort.

Its profile was classic. From the field, which was cleared of all obstructions, it presented to view a series of upward sloping earth planes, each of which protected a masonry work beyond; the glacis, which protected the scarp, terminated against a brick parapet wall of the covered way banquette; above this was seen the exterior slope of the ramparts.

Because of the remote locations of the works in the vicinity of New Orleans and Mobile, Bernard provided for citadels which were normally used as quarters. The plans of these buildings, sometimes called defensive barracks, were derived from the configurations of the spaces defined by the terrepleins and parade walls. Similar in concept to medieval keeps, these structures, with brick walls approximately four feet thick, were to be where the defenders could prolong resistance with a last stand, should an enemy carry the main body of the place. At Fort Morgan, for active defense, rifle embrasures were provided in three tiers, the upper one of which commanded the enceinte.

Several years after Bernard completed plans for the defense of Mobile Bay, the board of engineers was directed to project works for the western end of Santa Rosa Point to defend Pensacola Bay. The fortifications that had been ceded to the United States in 1821 would have offered but feeble resistance to any foe. Eventually named Fort Pickens, the work designed for the point by Bernard—the individual member of the board who was sent to do the planning—was begun in 1828 and, at the time, was the largest and most complex military structure on the Gulf (fig. 52). Although symmetrical in conception, the form recognized different exigencies of function arising from both bay defense and self-defense. Toward the water, guns were mounted in large casemates, two to each unit, and there was no plan for artillery above. The bastions oriented toward the land were the largest and most massive to withstand a potential siege; the one bastion and sides of two others which faced the water, which would be exposed only to the cannons of ships, were small and relatively light. Landward, cannons were mounted *en barbette* to defend against a siege, while guns for defense of the ditch were casemated.

WORKS IN THE FRENCH TRADITION WITH LARGE PERIMETERS

During the early part of the nineteenth century, then, the security of the Gulf Coast was largely entrusted to the fortifications designed by the talented Bernard. Although these were a great responsibility, his work in the United States was by no means limited to them. In addition to planning them

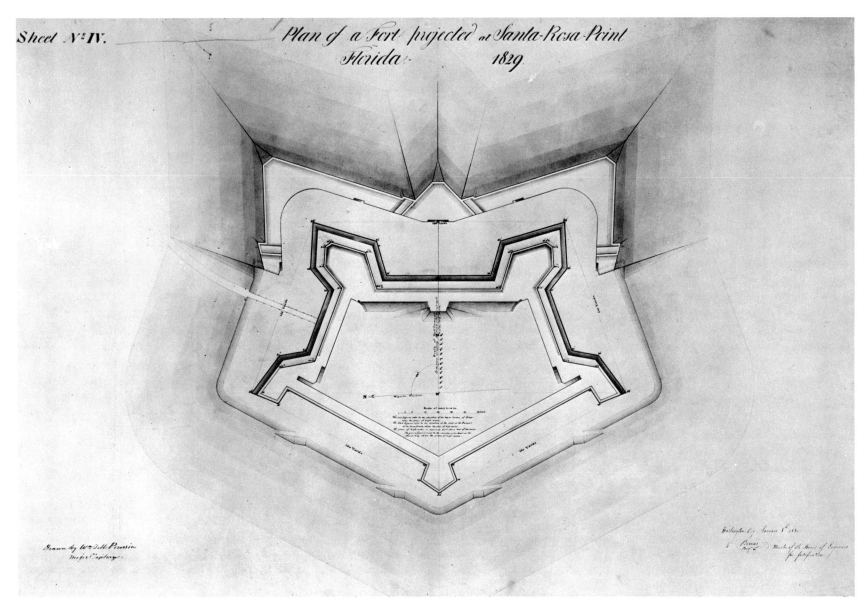

Plan of a Fort projected at Santa-Rosa-Point Florida. 1829

Drawn by Wm Tell Poussin
Major Engineer.

52. Fort Pickens, Florida (1828). Simon Bernard, engineer. Site plan (1829) drawn by Guillaume Tell Poussin. *National Archives, Washington, D.C.*

and playing a major role in the development of continental highways, he conceived other important forts along the Atlantic Coast.

In 1818, following his study of the Gulf, Bernard reconnoitered the Chesapeake Bay. Subsequently, for Old Point Comfort, he designed Fort Monroe, Virginia,[14] a first-class work commenced in 1819 and completed ca. 1847 (fig. 53). In area the largest work in the permanent system,[15] Fort Monroe was of sufficient magnitude to enable Bernard to eliminate weaknesses that were often inherent in works of small perimeter. Although early plans show it to have been an irregular work, the northwest bastion was moved outward to balance with the northeast bastion. As finally constructed, the fort was symmetrical, technically described as a regular bastioned work with seven fronts. It was planned for a peacetime garrison of 600 men, but it was thought that 2,625 men would have been needed to bring it up to strength sufficient to resist a siege.

Located on a point connected to the mainland by only a narrow isthmus and bridge, Fort Monroe was strong by virtue of its position and form; to take it by a regular siege would have been virtually impossible. Unlike his work along the Gulf, since it would have been difficult to attack by land, Bernard excluded from the original plans regular outworks—glacis and covered ways—intended to shield the enceinte from cannon fire. However, in front of the northeast curtain—the only one exposed to batteries which might have been located on the mainland—plans show a reentering place of arms, and in front of the adjacent bastions were parapets. A redoubt located in advance of the northern bastion defended the isthmus. Were an enemy successful in passing the redoubt or in landing on another section of island, he would have been confronted by a formidable deep wet ditch, or moat, which varied in width between 60 and 150 feet. The enclosing masonry wall at Fort Monroe, some ten feet thick at the base, was revetted with granite, as was the counterscarp, after the work on the main body of the trace was complete (fig. 54). Both granite and brick were used to face the parade wall in areas where the fort was casemated; in other areas the ground sloped up to the terreplein and was sodded.

Five years after work on Monroe was initiated, the second Fort Adams, Rhode Island (1824–ca. 1845), another formidable work, was begun on an irregular projection of land, replacing the work which had been designed by Louis de Tousard before the turn of the century (fig. 55). Since Adams was designed to defend what was considered to be the best harbor on the coast of the United States, it was conceived as one of the strongest works. In size and construction Fort Adams rivaled Monroe, but it exceeded Monroe in quantity of armament. In perimeter Fort Adams measured 1,739 yards compared to Monroe's 2,304. Adams was designed to mount 468 guns, while Monroe was planned for 380; the siege garrisons of both were more than twenty-four hundred men. Similar to the Old Point Comfort Fort, the work at Brenton's Point was casemated and was revetted with shale and granite. As with Monroe, its comparatively large size revealed its considered importance to national defense based on geography, on past experience in war, and on prognostication. In area, however, Adams covered only slightly more than one-third the ground occupied by Monroe.

The original plan of Fort Adams, also the concept of Simon Bernard, was entirely different from Fort Monroe and was perhaps the most complex ensemble of military architectural forms in the United States. The designer effected a configuration that was unique among forts in North America and, at the same time, a clear statement of form and function. According to J. G. Barnard, another eminent military engineer, Adams had no parallel at the time in the United States. Moreover, he wrote, "From its peculiar relations to the land defence, it

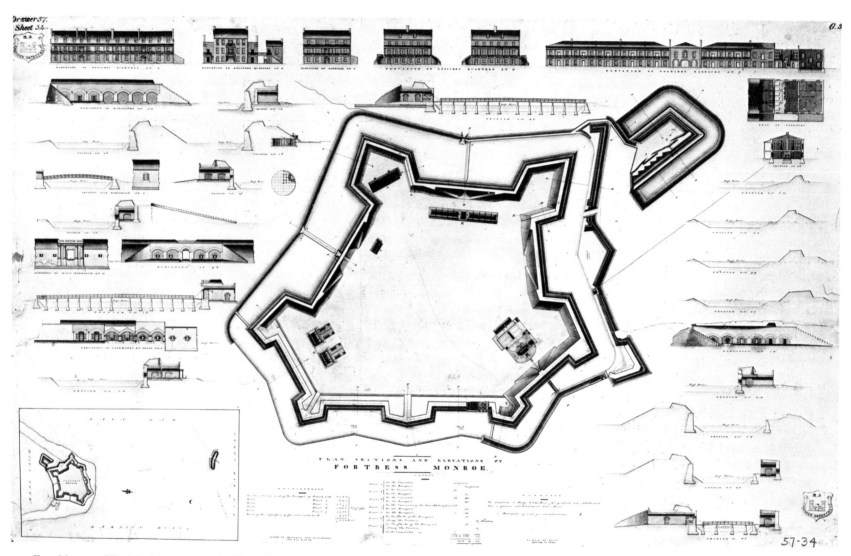

53. Fort Monroe, Virginia (1819–ca. 1847). Simon Bernard, engineer. Plan and sections of fort and elevations of buildings drawn by George O'Driscoll. *National Archives, Washington, D.C.*

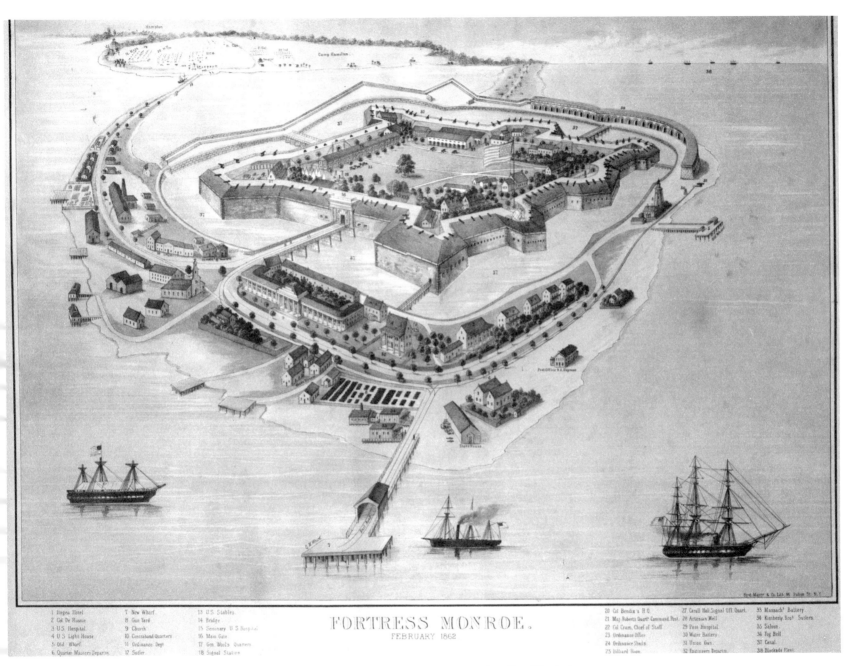

FORTRESS MONROE.
FEBRUARY 1862

1 Hygea Hotel	7 New Wharf.	13 U.S. Stables.	20 Col Bendix's H.Q.	27 Carell Hall Signal Off. Quart.	33 Massach. Battery
2 Col. De Russie	8 Gun Yard	14 Bridge	21 Maj. Roberts Quart. Command Post.	28 Artesian Well	34 Kimberly Bro. Sutlers
3 U.S. Hospital	9 Church	15 Seminary U S Hospital	22 Col Cram, Chief of Staff	29 Post Hospital	35 Saloon
4 U.S. Light House	10 Contraband Quarters	16 Main Gate	23 Ordinance Office	30 Water Battery	36 Fog Bell
5 Old Wharf	11 Ordinance Dept	17 Gen. Wool's Quarters	24 Ordinance Shein	31 Union Gun	37 Canal
6 Quarter Masters Departm	12 Sutler	18 Signal Station	25 Billiard Room	32 Engineers Departm.	38 Blockade Fleet

54. Fort Monroe, Virginia (1819–ca. 1847). Lithograph (1862). *Amon Carter Museum of Western Art, Fort Worth, Tex.*

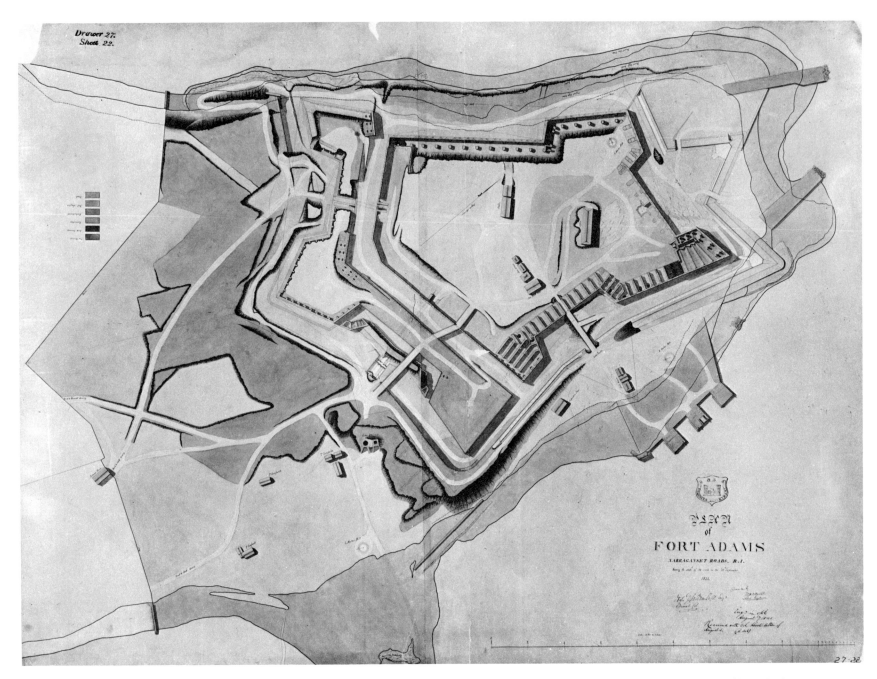

55. Fort Adams, Rhode Island (1824–ca. 1845). Simon Bernard, engineer. Plan of the state of the works on September 30, 1831. *National Archives, Washington, D.C.*

called for the application of most of those rules of art and many of those special arrangements which form the themes of Treatises upon 'fortification,' and which generally have but a very limited application to works of harbor defence."[16]

Owing to the shape of the land, the water fronts—fully casemated—were irregularly traced, consisting of a demibastion and two bastions with uneven faces connected by curtains. Comprising primarily the main body of the place, these features were designed for water defense only and were connected on the landward side by a long curtain, without casemates, forming an obtuse angle. Protecting the main body from the land approaches—where prolonged heavy bombardment was feared—were two other fronts with large full bastions completely separated from the main body by a wide ditch. The curtains of these fronts were protected by tenailles with casemated flanks—the concept for the use of the former in military architecture had been developed by Vauban; the latter was developed late in the eighteenth century by Henri Jean Baptiste Bousmard (1749–1807). Located between the flanks and before the curtain, these works protected the postern and the flanks along with the entire curtain (in the drawing showing the state of the works in 1831—which incorporated some changes made by J. G. Totten to the original trace—one of these tenailles had not yet been started).[17] Although rather complex, this architectural arrangement made the fort extremely strong against a land attack. If a tenaille or one of the landward bastions were breached, the enemy still would not have had access to the enceinte; at least two breaching operations would have been necessary.

Further increasing the resistance of Fort Adams to a regular siege by land was the system of countermining tunnels incorporated into the works—a system which had been frequently used in France and which, on a smaller scale, was also used at Fort Pickens. Designed to detect enemy efforts to tunnel under parts of the fort, plant explosives, and destroy entire sections of earth and masonry, this network consisted of counterscarp galleries, galleries of communication, and listening galleries. Located along the counterscarp of the interior ditch, the counterscarp galleries, accessible from the enceinte through two tunnels extending under the interior ditch, provided both cannon and musket reverse fire on the ditch and furnished communication between the galleries of communication extending under the bastions and under the exterior ditch to a gallery along the outermost counterscarp. From these galleries the listeners extended along the scarps and, on the southeast, outward under the glacis. Located at evenly spaced intervals of about twenty feet were openings, temporarily closed with stone panels, which would enable counterminers to detect the sounds of enemy miners' tools, to home in on the activity, and to intercept and destroy the tunnels. In addition to this main function, countermining tunnels were a psychological deterrent to any enemy; fearing the ground might be exploded by the besieged, aggressors were reluctant to attack over ground known to be countermined. In architectural form, size, and details, all of this design, along with a large redoubt to the south, made Adams one of the most sophisticated and significant bastioned forts on the continent.[18]

POLYGONAL FORTS

The bastioned system of fortification employed at Fort Adams and other early nineteenth-century strongholds had evolved in answer to the maxim, "All parts to enclose a space of ground, ought to be flanked, that is to say, viewed from every side, that there be no shelter about the place where the Enemy may lodge himself."[19] For close-in defense the best system was that which had the best flanking arrangement; the flanks of the bastioned system provided a view and coverage of opposite faces of the bastions and adjacent curtains, the latter of which were often casemated to assist in the coverage of the ditch. While this arrangement was extremely effective for large strongholds, it was sometimes difficult to avoid blind spots in small works. Small flanks, characteristic of little forts, were an additional weakness since they could contain little armament. They could not, in a small work, be increased in size without diminishing the length of the faces of the bastion and strangling the bastions at the gorge. Therefore, the bastioned mode was not appropriate for all forts.

There were, however, other methods of flanking. During the eighteenth century, counterfire rooms (embrasured galleries located behind the counterscarp) had been used for the defense of detached lunettes in France.[20] Although the French did not consider this a good system for forts with an extensive perimeter, it was suitable for small works. In addition, galleries located behind the counterscarp were less exposed to enemy artillery than were bastion flanks; this was one of the most significant advantages.

Fort Macon at Bogue Point, North Carolina (1826), another link in the permanent system, was a fine example of the use of counterfire rooms to flank the ditch (fig. 56). The ditch surrounding the small but aesthetically graceful five-sided brick enceinte was flanked and completely covered by four counterscarp galleries. Accessible either by doorways opening onto the ditch or by stairways descending from the covered way, these vaulted rooms were provided with embrasures for small cannons and loopholes for musketry.

Although the trace of Fort Macon was not common in America, the profile was consistent with well-established principles of the art of fortification. All that could be seen from the field were the earth glacis and ramparts. The sally-port was approached through a cut in the glacis, which was sharply curved to prevent the exposure of the gate to direct fire.

Another example of a form developed with counterfire rooms and influenced by the character of the terrain was Fort Barrancas, Florida (1839–44). A brick construction on a hill directly overlooking the Spanish battery, San Antonio (Fort San Carlos), and across the bay from Fort Pickens—a site which had been recommended by Captain James Gadsden (1788–1858) for a permanent work more than three decades earlier—it had a polygon of fortification in the shape of a trapezium (fig. 57).[21]

The ditch which paralleled two sides of the main body and the exposed faces of the enceinte overlooking the Spanish battery were flanked in each direction by cannons located in casemates behind the counterscarp wall. Continuous counterscarp galleries connecting these chambers and scarp galleries along the enceinte contained closely spaced rifle ports. Like the casemates, both were vaulted. Since the fort was situated on a rather steep hill, it was impossible and unnecessary to build a counterscarp on two of the sides.

Access to the rooms across the ditch from the main body of the place was through a subterranean tunnel. Other underground passages provided for communication with the Spanish battery and an advanced redoubt (1844), a brick casemated work located some fifteen hundred yards to the north to protect the land approach from that direction.

Built despite immense structural difficulties created by soft, unstable soil, Fort Livingston, located on a site controlling the pass into Barataria Bay, Louisiana, was similar to Fort Barrancas in form and size. Bernard's original plan showing bastions was completely rejected in favor of the polygonal concept, a simplification intended to reduce the cost.[22] The completed fort was a finely detailed work of brick incorporating reverse

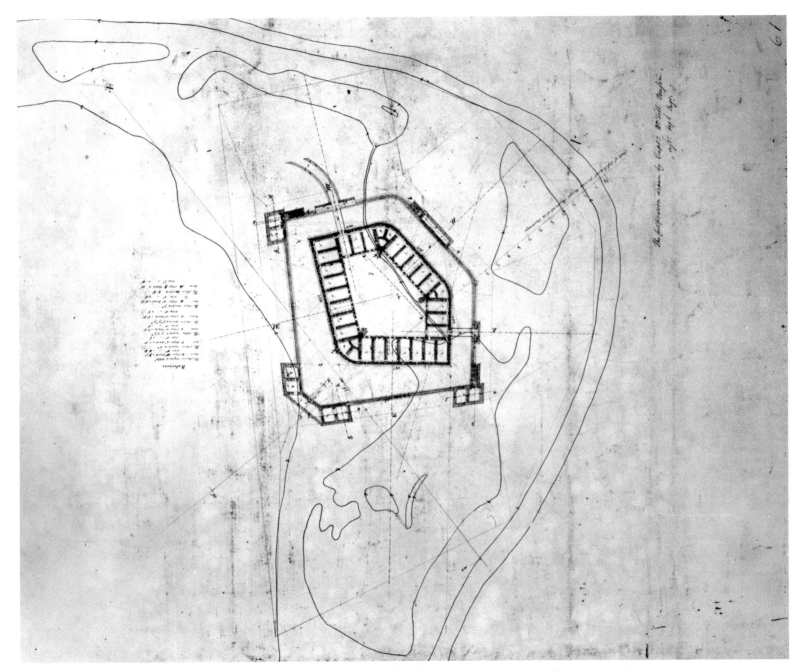

56. Fort Macon, North Carolina (1826). Plan (1821) drawn by Guillaume Tell Poussin. *National Archives, Washington, D.C.*

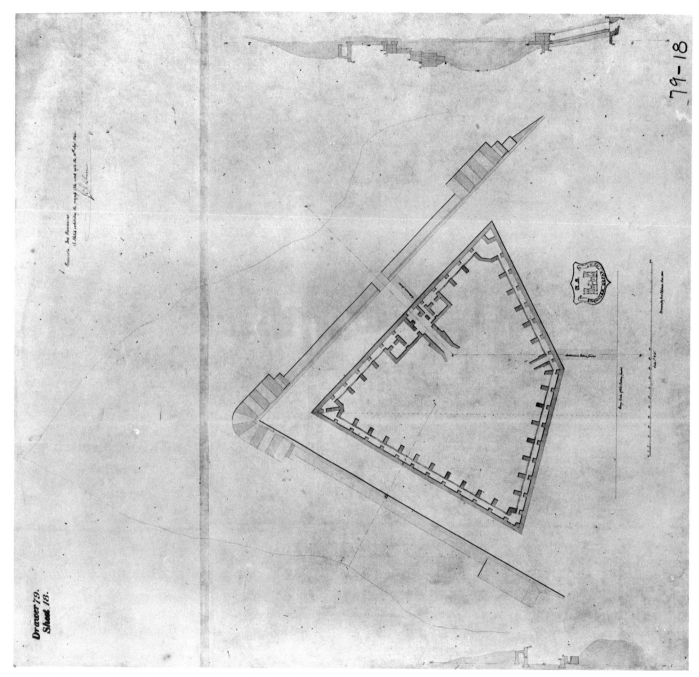

57. Fort Barrancas, Florida (1839–44). Plans and sections (1842). Progress drawing by George E. Chase. *National Archives, Washington, D.C.*

fire galleries for both cannons and muskets. As was usual, casemates of the main body of the work contained quarters and other essential services. Today, however, the fort is in ruins, having yielded to the forces of nature.

FORTS SUMTER AND PULASKI

Since the south Atlantic Coast was considered less important than some other areas of the permanent system, it was not until 1826 that the board of engineers prepared plans for the defense of Savannah and Charleston harbors. The forts designed for these two locations, possessing some similarities of trace, further epitomized forms that beautifully answered certain specific requisites of function. The functional requirements for the defense of these harbors, like the defense of those previously analyzed, were, however, subject to some variation according to the interpretations and judgment of designers and the site conditions.

After the board of engineers visited Charleston Harbor, they decided that a work of a permanent character was needed to create a crossfire with Fort Moultrie in order to adequately protect the commercial city of Charleston.[23] The main ship channel passed near Sullivan's Island but, fortunately for defense, was separated from Morris Island to the south, and it was narrowed by a long shoal. It was at the edge of this shoal, upon an island to be created, that it was determined to locate Fort Sumter, South Carolina (1829).

After careful analysis, plans were developed for a new brick fort with a five-sided polygonal trace (fig. 58). According to the report of the board, the relationship with the existing fort determined the orientation of the new work, while the defensive function determined the form and size.

The fire, of the projected work, will begin to take effect, upon approaching vessels at the same time with that of Fort Moultrie; &

from this moment it will be equallized upon all parts of the channel. . . .

The *form*, of the work in question, was determined by the space to be commanded. Giving to the guns a traverse of sixty degrees, four sides of a hexagon distribute the fire equally over all this space. The enclosure is completed by a straight gorge. The *magnitude* was the result of several considerations: first, . . . the value of the object to be defended . . . second, . . . the accommodation of the garrison, and proper service of the work.[24]

Since the fort was to be situated on a small mole, it was necessary to select a concept on vertical fortification in order to accommodate the required artillery (fig. 59). On all sides except the gorge there were two tiers of casemated guns, the embrasures for which were trimmed with granite, and one tier *en barbette* behind a brick parapet, which was trimmed on the exterior with a beautiful brick corbel table of arches. In the concept of casemates and multiple tiers of artillery it recalled the early nineteenth-century castles and the theories of Montalembert on seacoast defense. In the tiers contained within the gorge were magazines and quarters, the latter lighted by thin, vertical windows.

The geometry of the trace of Fort Pulaski, Georgia, commenced in 1829 and largely completed over two decades, was like that of Fort Sumter (fig. 60). Located near the edge of Cockspur Island for the defense of the Savannah River, its salient also pointed in the direction which provided the best coverage of the river approaches by the faces and flanks of the enceinte. Also like the Charleston Harbor fort, the gorge, or back side, contained quarters instead of casemates for cannons.

Lucidly revealing the situation of Fort Pulaski, its location on a natural island approximately a mile long required the inclusion of outworks and some variations of architectural form to provide for close defense if an enemy should ever

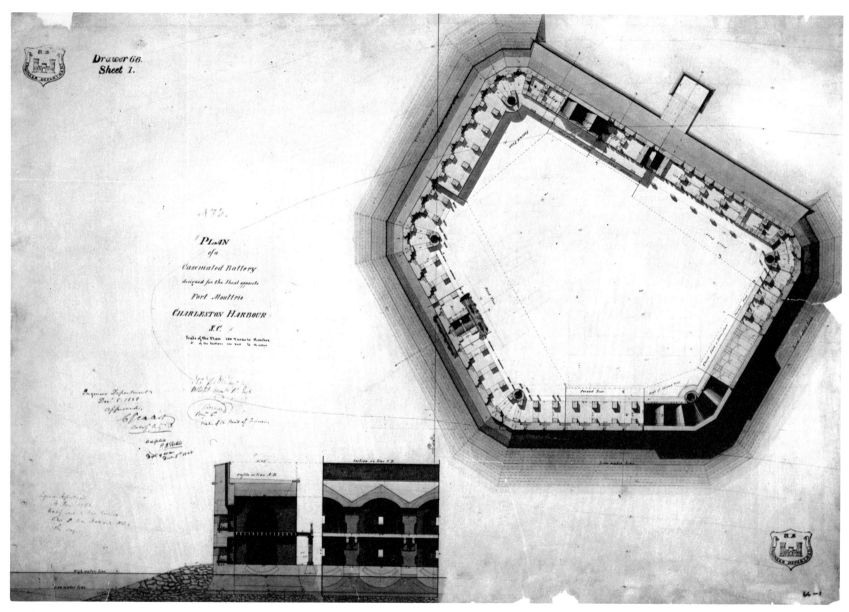

58. Fort Sumter, South Carolina (1829). Simon Bernard and Joseph G. Totten, engineers. Plan and sections (1828). *National Archives, Washington, D.C.*

59. Fort Sumter, South Carolina (1829). Painting (ca. 1870) by Seth Eastman. *Architect of the Capitol, Washington, D.C.*

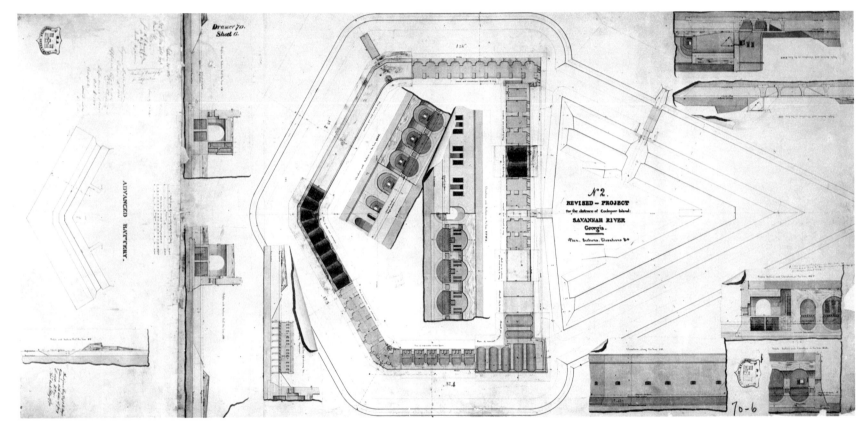

60. Fort Pulaski, Georgia (1829). Simon Bernard and Joseph G. Totten, engineers. Plan, section, and elevations. *National Archives, Washington, D.C.*

secure a landing. A large ravelin, with earth terreplein and rampart, was placed to protect the gorge, which faced the landward side of the island. Separating this ravelin from the main body and completely surrounding both was a wide wet ditch, the water for which was obtained by canal from the river and controlled by sluice gates. Considerable attention was given to the defense of the gorge and its ditch. The gate, located in the center, was secured by a drawbridge and portcullis that rose and lowered synchronously (fig. 61), and a set of heavy double doors was located inside the portcullis. As added security for this side of the work, demibastions were provided, the flanks of which contained an embrasure and two loopholes looking along the curtain. The faces of these demibastions each had another embrasure and four loopholes directed down the ravelin ditch.

The architectural form for Fort Pulaski is one of the clearest and handsomest expressions of structural function to be found in a North American fortification. To spread the weight of massive brick walls, which reached twenty-five feet above water, and the vault-supported terreplein of the second tier, the base of the exterior wall was boldly splayed out to the edge of a timber grillage. On the interior, the weight of the vaults and countervaults was transferred evenly to the grillage by wide reversed arches. To reduce the weight to be supported on the soft soil, the floor under the casemated tier of guns was constructed of wood.

TOWER-BASTIONED FORTS

In the 1830s and 1840s controversies developed between military engineers and congressmen over the magnitude of Forts Sumter and Pulaski as well as other forts in the permanent system.[25] However, advocates of the permanent system convinced Congress that formidable works were positively needed, and money continued to be appropriated for forts that were in progress and for new works. Perhaps debates over national defense proved more than before the importance of permanent defenses; at least the works subsequently undertaken architecturally were the most highly developed in form and construction. In the thirties and forties new variations of form evolved as a result of the search for effective and efficient arrangements of components in the forms designed to guard important positions along the coast.

Locations for these forts, as for their predecessors, were selected with regard to physiography and the geography of the national system, not with respect to desirable geological conditions. Consequently, engineers were often forced to work with difficult site situations. Soil conditions frequently presented difficulties in developing sound structural systems. Then, too, sites were most often restricted in area, necessitating several tiers of casemates in order to provide for the required amount of artillery. Problems arising from these circumstances were encountered at Fort Delaware (1818–26), a pentagonal casemated work on Pea Patch Island in the Delaware River. In 1833 it was demolished because of large cracks caused by unsound foundations. To protect the city of Philadelphia, it was replaced by a new, permanent fort (1836–59) mounting three tiers of guns on five sides (fig. 62). The massive walls of the new work were supported on a network of some twelve thousand piles driven into the mud of the Pea Patch.

Among the other points that concerned engineers at this time were Key West and Dry Tortugas off the southwestern Florida Coast. The good harbors of these positions, it was feared, might provide an adversary safe resorts and bases from which he might direct attacks against the mainland. Fortifications were required in order to deny their occupation.

In an 1840 report, Joseph G. Totten, then chief engineer, broadly sketched the determinants of design for the forts at these locations. "The works must be adequate to resist escalade, bombardment, and cannonade from vessels, and to

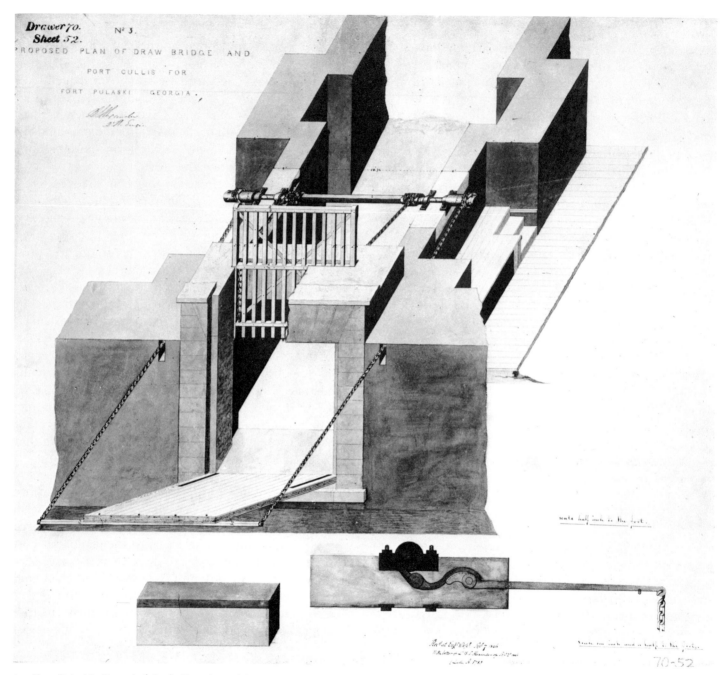

61. Fort Pulaski, Georgia (1829). Drawing of drawbridge and portcullis. *National Archives, Washington, D.C.*

62. Fort Delaware, Delaware (1836). R. Delafield, engineer. Painting (ca. 1870) by Seth Eastman. *Architect of the Capitol, Washington, D.C.*

63. Fort Taylor, Florida (1845). Joseph G. Totten, engineer. Painting (ca. 1870) by Seth Eastman. *Architect of the Capitol, Washington, D.C.*

sustain a protracted investment; but . . . they will be not exposed to any operations resembling a siege. . . . They must have capacious store-rooms, be thoroughly bomb-proof, and be heavily armed."[26]

Since these forts were also built on confined sites, they were, like Sumter and Pulaski, multitiered, casemated works and were based on regular polygons of fortification. Near Key West, Florida, Fort Taylor (1845) was a large granite and brick structure of two tiers with a plan developed on a polygon of a half hexagon (fig. 63). A significant modification, which had already appeared in the design of Fort Delaware, was the addition of small, wide-flanked tower bastions designed to contain howitzers to enfilade the curtains to prevent escalade—forms which recall the tower bastions of Vauban's second and third systems of fortification. Compared to the bastions on land fronts, these tower works were quite small in area, since they would not be exposed to the operations of a regular siege.

Tower bastions also characterized Fort Jefferson, Florida (1849), located on Garden Key of the Tortugas some sixty miles east of Fort Taylor to strengthen the security of the Gulf (figs. 64, 65). Evidently designed by Joseph G. Totten, the plan was developed on a six-sided polygon adjusted to the irregular coastline of the island and with a narrow flanking work at each point. Typical of American seacoast forts built during this epoch, the width of the flanks of these tower bastions exceeded the length of the faces of the bastions.

Among the largest fortifications to be undertaken in the permanent system, Fort Jefferson enclosed most of the sixteen-acre key and had a perimeter circuit of approximately one-half mile. It was designed to garrison fifteen hundred men and to mount 450 pieces of artillery in three tiers, although the embrasures for the second tier were never completed. In each of the first two tiers there were 36 casemates in the bastions and 122 casemates in the curtains, 111 of which on the ground level contained artillery. In other casemates on the bottom tier were a sallyport, a magazine, and guardrooms.

Other smaller forts commenced during the 1840s possessed similar characteristics of trace but were modified to adapt to their position. Now in the shadows of the awesome Verrazano Narrows suspension bridge, Fort Richmond (Fort Wadsworth), New York (1847), had a polygonal form which, like that of Fort Taylor, was one-half of a hexagon (figs. 66, 67).[27] Richmond was designed as a companion to Fort Hamilton (1824) across the narrows, and the two angles of the polygon facing the water were provided with tower bastions similar to those of the Florida forts. However, on the angles of the land side of the four-tiered stone fort were regular bastions with uneven faces.

Works with Detached Scarps

After improving the new permanent seacoast forts with the addition of tower bastions, other developments occurred based on innovations that were made in France. As in earlier periods, throughout the first decades of the nineteenth century military architecture in the United States was based on French theory, considered to be the paragon of the art; the United States was never free from that authority.[28] Simon Bernard and other French engineers who preceded him applied in America many of the principles followed by Vauban and his successors. Further ingraining the French tradition in military architecture, instruction in fortification at the United States Military Academy in the early years conformed to theories developed in France.

The principles of profiling in the French system of fortification, as embodied in the early works of Bernard and other French engineers, were well proven and subject to less ex-

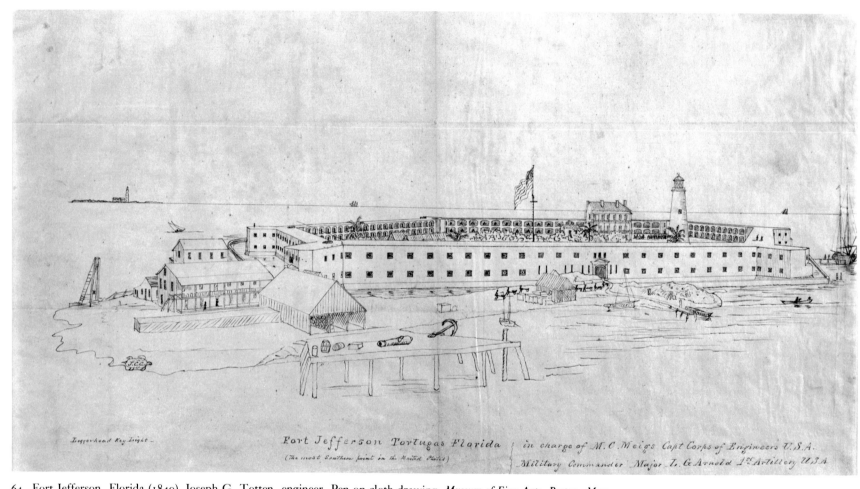

Loggerhead Key Light —

Fort Jefferson Tortugas Florida
(The most Southern point in the United States)
in charge of M. C. Meigs Capt Corps of Engineers U.S.A.
Military Commander Major L. G. Arnold 1st Artillery U.S.A.

64. Fort Jefferson, Florida (1849). Joseph G. Totten, engineer. Pen on cloth drawing. *Museum of Fine Arts, Boston, Mass.*

65. Fort Jefferson, Florida (1849). Painting (ca. 1870) by Seth Eastman. *Architect of the Capitol, Washington, D.C.*

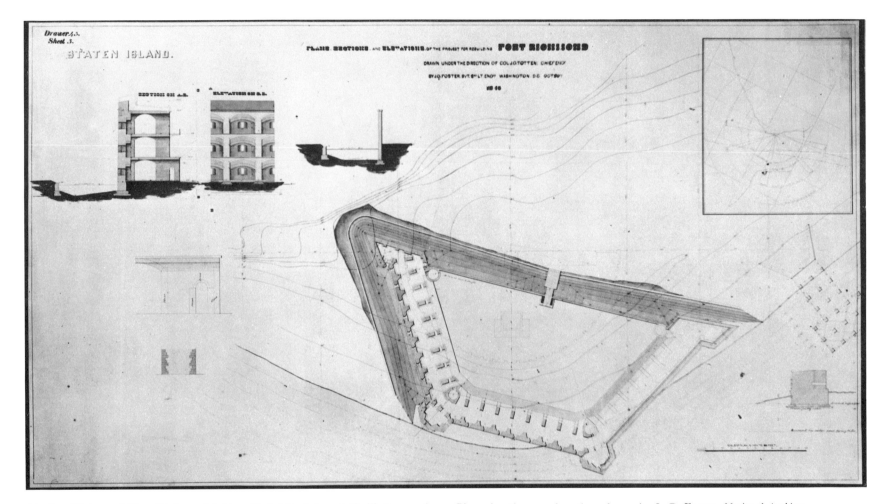

66. Fort Richmond (Fort Wadsworth), New York (1847). Joseph G. Totten, engineer. Plan, elevations, and sections drawn by J. G. Foster. *National Archives, Washington, D.C.*

67. Fort Richmond (Fort Wadsworth) and Fort Tompkins, New York (1858). Painting (ca. 1870) by Seth Eastman. *Architect of the Capitol, Washington, D.C.*

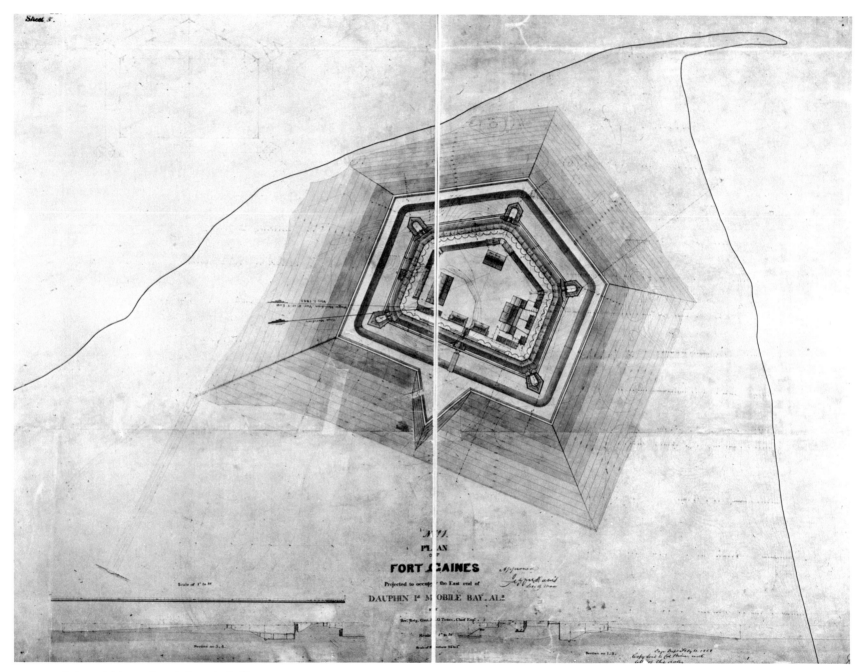

N.º 1.

PLAN
OF
FORT GAINES

Projected to occupy the East end of
DAUPHIN I.ᵈ MOBILE BAY, AL.ᵃ

68. Fort Gaines, Alabama (1854). Joseph G. Totten, engineer. Plan and profiles. *National Archives, Washington, D.C.*

perimentation than the plan arrangements of various architectural defense components. However, in the first half of the nineteenth century new works at Lyons, detached forts of Paris, and modifications of old works appeared with profiles characterized by detached scarps,[29] which were originated by the eminent Frenchman Lazare Nicolas Marguerite Carnot (1753–1823).[30] In this new system, the scarp was moved away from the ramparts, the exterior slope of the latter terminating at a *chemin de ronde* on the inside of the wall considerably below the level of the cordon. It was believed that this arrangement created a more stable rampart. When battered down by artillery, the scarps would not bring down with them the earth of the rampart, which would then create a ramp over which the enemy could storm the main body of the place. Carnot replaced the steep masonry counterscarp which traditionally surrounded the enceinte and defined the outer edge of the ditch with a gentle earth slope. This, he advocated, facilitated sorties against besiegers, actions considered to be very important in successful defense. From these developments it was apparent that the traditional influence of Vauban had waned in France and, subsequently, in America.

The "Carnot wall," following the lead of the French, appeared in two forts in the United States designed in the 1840s: Fort Clinch, Florida (1851), for the defense of Cumberland Sound, and Fort Gaines, Alabama (1854),[31] across from Fort Morgan (fig. 68). Conceived by Joseph G. Totten, these two works were similar in size, plan, and section, since their functions were virtually identical. In form they represented a synthesis between the polygonal plan type with tower bastions, then in use in America, and the new profile with a detached scarp that had been developed in France.

The configuration of both of these forts reveals strong separation of two distinct defensive functions, the nature of which was set forth in an 1840 report on fortifications: "... fortifications of the coast must ... be competent to the double task of interdicting the passage of ships and resisting land attacks—two distinct and independent qualities. The first demands merely an array, in suitable numbers and in proper proportions, of heavy guns, covered by parapets proof against shot and shells; the second demands inaccessibility. ... there is nothing in the first quality necessarily involving the last."[32]

In the five-sided plans of Forts Clinch and Gaines, cannons for defense of the water were mounted *en barbette* behind brick walls fronted with earth parapets and were concentrated on the face and flank curtains. Separated from the ramparts by the *chemin de ronde*, the scarps were designed to fulfill the function of inaccessibility—to repel attempts at sudden escalade or any other efforts to seize the places by land. Along these curtains were rifle ports, the checks of which were formed with a series of offsets instead of smooth, slanted planes to avoid deflecting enemy bullets through the throat. In conformance with Carnot's theories, the scarps were protected by glacis, and the counterscarps were replaced with gentle earth slopes.

Also indirectly reflecting considerations resulting from the two separate defensive functions were the distinct traces of the flanking works. The gorge bastions themselves had narrow gorges, while the salient and shoulder bastions (tower bastions) had wide gorges with parallel flanks. Located on the landward side, where the sides of the polygon of fortification formed an angle of ninety degrees, the distance between the flanks of the gorge bastions diminished inward to create satisfactory lines of fire along the curtains. The interiors of all the bastions were quickly accessible through tunnels under the terreplein and ramparts or directly from the *chemin de ronde* (fig. 69). Fortifications similar to Gaines and Clinch were pro-

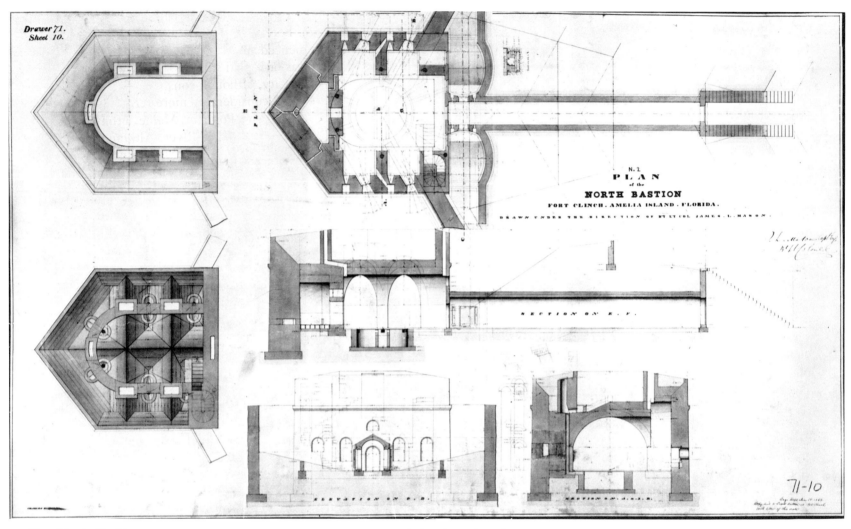

N. 2
PLAN
of the
NORTH BASTION
FORT CLINCH. AMELIA ISLAND. FLORIDA.

DRAWN UNDER THE DIRECTION OF BT LT COL JAMES L MASON.

SECTION ON E F.

ELEVATION ON C D.

SECTION ON A B.

71-10

69. Fort Clinch, Florida (1851). Joseph G. Totten, engineer. Plan, sections, and elevations of north bastion drawn by Charles Haskins. *National Archives, Washington, D.C.*

jected for Galveston Harbor, Texas, in 1859 but were never realized.

END OF AN ERA

National defense, although concentrated along the coasts of the Atlantic Ocean and the Gulf of Mexico, was not confined to those areas alone. Along the northern frontier, Fort Wayne, Michigan (1842), a square, four-bastioned work with a brick scarp and casemates, was constructed in response to tensions between the United States and Canada; Fort Ontario, New York, was rebuilt (1839) with stone and earth on a pentagonal, bastioned trace; Fort Niagara, New York, was maintained, and as late as the Civil War brick casemates were added. To protect against any possible invasion by the British via Lake Champlain, work was commenced in the 1840s on Fort Montgomery, New York, considered to be the most important structure along the border since it defended one of the great natural highways of the continent.

Upon the far westward extension of the American frontier, plans were developed in the 1850s for the defense of key bays of the Pacific. San Francisco Bay was fortified with works on Alcatraz Island (1854) and at Fort Point (1853) on the south side of the Golden Gate on the former site of Castillo de San Joaquin, a Spanish work. At Alcatraz, defensive barracks crowned the rugged island while miscellaneous batteries were located along the perimeter; Fort Point, now dwarfed by the Golden Gate Bridge, was an irregularly traced, four-tiered brick and granite work which could mount 126 guns. These were the most formidable works built on the West Coast.

Construction on multitiered, casemated forts continued past mid-century until the Civil War. Not until 1854 was construction completed on Fort Trumbull, Connecticut, a permanent work commenced in 1839 on the site of eighteenth-century fortifications (fig. 70). Fort Massachusetts, Mississippi, a controversial work commenced in 1859 but not completed until after the Civil War, was on a plan similar to that of the earlier castles near New York City, although construction was more massive and workmanship considerably more refined. As late as 1861 work was begun on Fort Popham, Maine, a third-class work situated to defend the Kennebec River. Although never completed, Popham was a three-tiered fort with walls of granite and vaults of brick. Occupying a rocky point, the gorge, flanked by demibastions, was straight while the front was parabolic in form. In the front the Totten casemate embrasures—named after the chief engineer who reduced the dimensions of the throat and external openings to considerably less than was used in Europe—were closed with iron shutters, a development which originated in the late 1850s to protect against grapeshot. It was a small fort and, unlike most other works begun in this period, had no tower bastions.

On a much smaller scale, the vertical concept of military architecture was used also in the design of several towers. At Proctor's Landing, Louisiana (fig. 71), and Key West, Florida, square towers three stories high were constructed with bricks. The work at Proctor's Landing was begun about 1856, and two towers at Key West were begun about 1860. Although they were sometimes called Martello towers—round or elliptical planned works with several stories[33]—they were actually modeled after the *tours-modèles* which were favored by Napoleon for coast batteries.[34] The thick walls of the first and second floors had loopholes for musketry, while on the top platform cannons were mounted *en barbette* behind masonry parapets. Within were magazines and quarters for the garrison.

As had always been the case, these last works, in configuration and magnitude, were developed following careful analysis of the role of each in national defense. Each fort, like the entire

70. Fort Trumbull, Connecticut (1839–54). Painting (ca. 1870) by Seth Eastman. *Architect of the Capitol, Washington, D.C.*

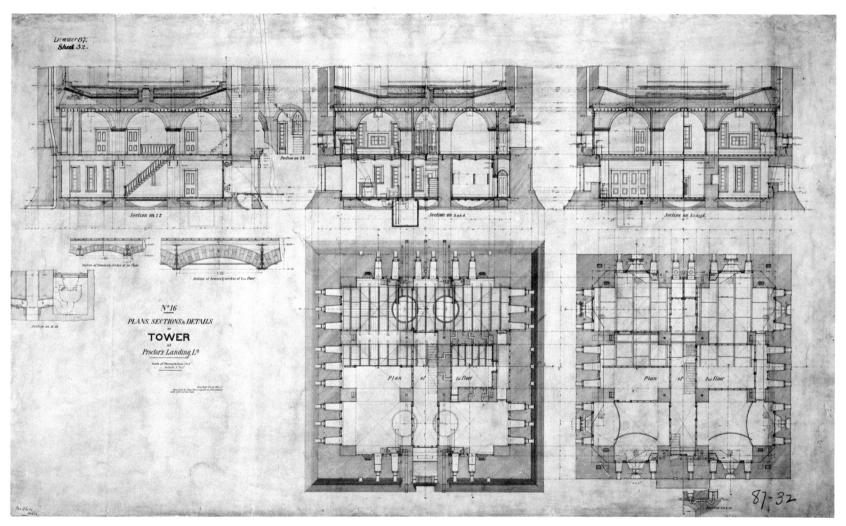

71. Proctor's Landing, Louisiana (1856). Plans, sections, and details. *National Archives, Washington, D.C.*

system, was designed according to analytical theory and by virtue of its form was conceived to be a strong work. All were logical expressions that were the outgrowth of national as well as local military purpose.

REVOLUTION IN FORTIFICATION

Ironically, the permanent forts of the eastern and southern coasts of the United States constructed after the War of 1812 never had an angry shot fired at them by a foreign power. The test of some, however, came during the Civil War.

Like the United States before the war, the Confederacy placed considerable confidence in the capacity of military architecture to defend its cities. The occupation of southern forts, it was thought, would be of moment to the success of the South. Immediately after the formation of the government, the Confederacy resolved to seize all the forts located in the southern states. Although the Stars and Stripes were never lowered at Forts Pickens, Taylor, Jefferson, and Monroe, Fort Sumter was occupied by force; others, having insignificant or no garrisons, were subsequently occupied without conflict.

Meanwhile, the war strategy of the North had been developed and included the blockade of the southern coast and the capture of the coastal forts which had been occupied by the Confederates. Pursuant to this policy, Fort Pulaski, considered to be one of the strongest forts along the coast, was attacked early in 1862. By virtue of its massive walls and the surrounding natural barriers, it was not generally believed that it could be taken except by investment. "The broad waters of the Savannah River and wide swampy marshes surrounded the fort on all sides. Ships of the Navy could not safely come within effective range of this citadel, and there was no firm ground on which land batteries could be erected nearer than Tybee Island, from 1 to 2½ miles away."[35]

Since the light, smoothbore cannons that were in use at the time Fort Pulaski was designed were only effective within a range of one thousand yards, the stronghold was considered secure. However, in the interval between the time that the fort had been designed and the Civil War new developments had occurred in weapons. The smoothbore Columbiads, which fired explosive shells ranging in size from eight to fifteen inches in diameter, and the massive Rodmans, which fired missiles ranging in diameter from eight to twenty inches, were developed. Then came the James and Parrott rifles, capable of firing with amazing accuracy at much greater range than the smoothbores. These weapons ultimately proved to be the nemesis of the art of permanent fortification.

Federal artillery commenced firing on Fort Pulaski on April 10, 1862. On the afternoon of the second day, the thick walls of Pulaski, unprotected by a glacis, were breached at the southeast shoulder angle. The breach was effected by light James and Parrott rifles assisted by eight- and ten-inch Columbiads.[36] Additional cannon fire was directed through the thirty-foot-wide breach diagonally across the parade toward the powder magazine at the northwest corner, and the fort surrendered. At Fort Pulaski, the destruction which formerly took many days—sometimes several weeks—to accomplish after siege batteries were constructed took less than two days from a distance nearly twice the effective range of light smoothbores.[37]

Hitherto considered with skepticism by officials, rifled cannons had virtually made obsolete all the forts that had been a part of the permanent system; fortifications based on theory that had taken centuries to develop no longer appeared adequate. Since walls of masonry could not long withstand the terrific impact of rifled cannons, the effect of these weapons on the architecture of forts in North America was to be as revolutionary as the invention of smoothbore cannons had

72. Fort Macon, North Carolina (1826). Interior and exterior of Fort Macon after the battle (1862); drawing by J. H. Schell from *Frank Leslie's Illustrated Newspaper*.

73. Fort Sumter, South Carolina (1829). Lithograph (1863). *Library of Congress, Washington, D.C.*

been centuries earlier in Europe. Nevertheless, it was only after several more sieges throughout the war that this fact was fully appreciated.

Shortly after the capitulation of Fort Pulaski, other forts formerly in the permanent system fell. Fort Macon was surrendered after an eleven-hour bombardment in which 560 of the 1,100 shots fired found their mark (fig. 72). Elsewhere, Forts Jackson and Saint Philip also fell, after sustaining considerable destruction, thus exposing New Orleans. It ap-

peared that there was no way the seacoast forts could withstand attack without major remodeling.

In 1863 important concepts of defense were learned and demonstrated at Fort Sumter. In 1861, after South Carolina seceded from the Union, Major Robert Anderson moved his blue-coated troops from Fort Moultrie to Sumter. Several months after the transfer, Confederates opened artillery fire on the fort, and Anderson surrendered after a short resistance, during which time fire and shot had destroyed sections of the work. The Confederates had immediately begun making preparations for a Union attack that was sure to come, since Charleston Harbor was a gap in the Northern blockade. Although badly littered, Sumter was yet structurally sound. Noting weak points revealed by the 1861 attack, engineers directed repairs and modifications. Some of the casemates were filled with sand, while other spaces were filled with cotton bales and sand. Then the gorge wall was faced with sandbags for two-thirds of its height.[38] All this improvisation was to strengthen the fort against rifles.

In the summer of 1863, Union batteries of rifled cannons were constructed on Morris Island, well out of range of Sumter's smoothbores. In August, Fort Sumter experienced its second and final siege. By delivering projectiles weighing 80, 200, and 250 pounds, Whitworth and Parrott rifles reduced the fort to an island of broken brickwork and debris in less than a week.

But the Confederates refused to surrender. Working at night, they defiantly and ingeniously rearranged the ruins, fashioning shelter in remaining casemates by covering them with masses of heavy timbers, sandbags, and earth. Whenever Union batteries fired at the fort, the determined Confederates filled in damaged areas with more sand or earth until it became essentially an earthwork (fig. 73). Even during the great bombardment of July, 1864, the Confederates held the work. With each repair and addition the fort appeared to get stronger, since even rifled cannon could not penetrate the masses of debris and earth—soft material was better than hard masonry. Fort Sumter was finally evacuated early in 1865, but then as a consequence of Sherman's march through South Carolina, not as a final consequence of Union cannon fire.

The final scene of encounter over a fort in the permanent system was Alabama. There, Confererate forces at the wane of the rebellion undertook the defense of Mobile Bay. But the fate of the forts at the entrance to the bay was the same as that of Forts Pulaski, Macon, and Jackson. Fort Gaines surrendered after three days of naval and land bombardment. Although Confederate engineers vainly attempted to strengthen Fort Morgan by erecting heavy sandbag traverses, completing the sand glacis, and reinforcing the magazines,[39] a combined land and sea attack overpowered it. After resisting a regular siege for two weeks, during which time several breaches were made, the commanding officer, General Richard L. Page (1807–1901), surrendered.

CIVIL WAR FORTIFICATIONS

At the beginning of the war, Confederate engineers, like those of the Union certainly familiar with new weapons but probably not suspecting their ultimate destructive capacity, followed traditional theory and practice on fortification. Along a newly established border time was critical; therefore, expediency was important. Rightfully, without regard to permanence, new works were built from earth and wood—not unlike many fortifications which were the products of previous conflicts.

The Confederate program for defense was extensive. In fact, it seems somewhat remarkable that so much was accomplished in so little time. In addition to new fortifications at the perimeter of the land frontier in the north and west and around key cities, new works were built along the maritime frontier and situated to protect lines of communication. Like works constructed inland, the new seacoast defenses were fashioned by rearranging the earth. Although they were easily damaged by erosion, they could be rapidly constructed and required no skilled laborers. Only the engineer in charge of the work and his assistants needed to be well trained to locate, trace, and direct the profiling of the fortifications.

Throughout the duration of the war Confederate engineers steadily improved the design of their coastal fortifications. Because of greater experience, they surely surpassed Union engineers in this specialty. Recognizing weaknesses of forts in the old permanent system and learning from numerous contests with Federal ships, they drew on their ingenuity to progressively alter and modify forms to increase resistance to naval attacks. After suffering many losses from enfilade fire on forts with continuous rows of guns mounted *en barbette*, cannons were separated by large, high traverses, thereby limiting the grazing destruction of missiles. When it was found that Union gunboats could, with concentrated fire, drive defenders from the closely spaced cannons on the terrepleins of the permanent forts, engineers adopted the practice of spacing batteries widely. At the end of the war, not less than sixty feet separated guns. Thus, late Confederate coastal forts presented lengthy fronts to besiegers.

Union engineers also designed a large number of new structures for defense. These, like those of the South, were impermanent and usually, by necessity, hastily constructed. Because of their irregular form, some appear to have been intuitive designs which responded to the form of the terrain and the estimated urgency of their construction. Others were carefully designed and traced in conformance to the precise geometrical configurations dictated by theory.

French influence in the art of siege and defense appears to have continued in Federal engineering during the Civil War. The technique of approaching Confederate works was essentially the same as that used by Vauban. In many of the larger earthworks, if genuinely designed to function as forts instead of batteries, bastions were employed as flanking components in conformance with the French school. Works too small for bastions were usually given the form of a simple polygon.

Among the early defenses by engineers of the North was a work (1861) at Fort Union, New Mexico. Near the post buildings established for the control of Indians, this fort was an earthwork with a ditch and earth parapet of strong profile thrown up in preparation for an attack expected from troops who had defected from the Union army. Incorrectly called a star fort, the trace was conventional, with a rectangular polygon of fortification and bastions at each corner.

Several Federal and Confederate towns which were, because of their location, industrially or politically important to the war effort were completely surrounded with works for defense. In designing defenses for cities, the philosophies of Union and Confederate engineers were essentially the same. Redoubts, lunettes, batteries, and forts were located around the perimeter. The terrain was always carefully analyzed, and works for defense were carefully located on rises to control all approaches. Depending on the nature of the ground and the planned defense when attack threatened, these fortified points were complemented with connecting trenches or other field fortifications.

Representative of the fortified cities was Washington, D.C. It was twice threatened—in 1862 and 1864. The possibility of attacks on the capital of the United States was ever on the minds of Union engineers throughout most of the war. On the very threshold of hostilities, the city was guarded only against naval advances on the south by Fort Washington. Like all the cities of the North and South, it was defenseless against a land-based attack. Throughout the duration of the war, however, it was progressively strengthened until it became one of the strongest cities on either side.

It was at the beginning of the war, under the direction of

military engineer John G. Barnard, that work was commenced on the cordon of defense around Washington. It was mandatory that roads to the city be protected, that the Potomac River approach be reinforced, and that the commanding heights on the Virginia side of the river be occupied with strong fortifications.

The first forts around the capital—Runyon, Corcoran, and Ellsworth—were in the Arlington-Alexandria vicinity. Located within the boundaries of the District of Columbia, all were begun in 1861 and were designed as semipermanent works. Fort Runyon, with a perimeter of 1,484 yards, was the largest. When this undertaking and several others were well advanced, works were initiated on the other side of the city, still in the district, to protect the main roads north. Fort Lincoln (1861) was located near the Baltimore and Ohio Railroad, Fort Stevens (1861) was situated near the Seventh Street Road, and Fort Reno (1861), the largest of those north of the Potomac, guarded the Rockville Road. After providing for the defense of key points, the areas in between were filled with batteries and smaller forts, spaced apart at distances less than the range of cannon fire. By April, 1865, the capital had been surrounded with sixty-eight forts and batteries having an aggregate perimeter of some thirteen miles; there were 35,711 yards of entrenchments, and the entire system surrounding the city was thirty-seven miles long.[40]

Influence of the Civil War on Fortification

Out of the Civil War evolved new concepts of fortification. Many of these were the antithesis of previous theories and involved changes in construction materials, revisions of functional arrangements, and reevaluations of the capabilities of permanent defenses.

The foreign engineer Viktor Ernst Karl Rudolph von Scheliha, who served with the South at Mobile, set forth several new principles of coast defense based on personal experience and supported by official reports made by Federal naval officers who had attacked works defended by Confederates. Relative to military architecture, von Scheliha advocated the following principles:

Exposed Masonry is incapable of withstanding the fire of Modern Artillery.[41]

Earth, especially Sand-works, properly constructed, [is] *better Protection against Modern Artillery than the permanent Fortifications built on the old plan.*[42]

No Forts now built can keep out a large Fleet unless the Channel is obstructed.[43]

Immediately after the Civil War few new structures for defense appeared; forts damaged during the war were restored, incomplete works were finished, and other works were kept in ordinary repair or were partially modified. In the 1870s the board of engineers prepared plans to modify existing forts by installing earthen batteries with large calibre artillery, river obstructions, torpedoes, and floating batteries. Although work was begun on many of the proposed plans, appropriations were withheld and modifications on some were never completed; on others they were never begun. Meanwhile, despite some alterations to existing forts, most critics considered key urban areas virtually unprotected. Exposed and defenseless American cities were subjects of wide popular and official discussions and criticism.[44] Perhaps because no one had satisfactory answers to the problem and because of rapid developments in increasing the seemingly endless power of artillery, Congress refused to appropriate money for new works.

In 1885 a group known as the Endicott Board was formed to cooperate with the board of engineers in developing a new system of defense.[45] Plans were developed for fortifications which included armaments of the heaviest rifles mounted on disappearing carriages,[46] a system of submarine mines, and batteries of rapid-fire guns of small calibre to protect mined areas, all of which were to be protected by massive works of reinforced concrete. With appropriations made in 1890 construction was begun on works for the protection of New York, Boston, Washington, Hampton Roads, and San Francisco, the cities where fortifications were considered most urgent. In subsequent years other cities including Newport, Mobile, and New Orleans were also fortified. Many of these new works were physically incorporated into the old forts of the permanent system. At Fort Morgan a concrete battery (1898) was located across the parade; at Fort Gaines batteries (1901) were located behind the original earth parapets, and the *chemin de ronde* was filled with sand; at Fort Sumter, which had been partially rebuilt in the 1870s to mount large Parrott rifles and Rodman smoothbores, a battery (1899) was constructed in the middle of the parade and the area between the casemates and the battery was filled with sand. Likewise, the parade at Fort Pickens was the location of a new battery (ca. 1898), the southern casemates being filled with sand to shield it.

Although attempts were made to develop stronger fortifications through modification and innovation, architecture for defense was simply unable to keep pace with weapon design. Nonetheless, the active military history of some of the permanent forts did not end until World War II, when they were permanently disarmed. Although their role has changed, the history of some is yet incomplete.

4 Land Frontier Forts

During the period roughly encompassing the last quarter of the eighteenth century and the first half of the nineteenth, the United States, by settlement, purchase, treaty, or annexation, vastly extended her boundaries. The furs, agriculture, and minerals that came with new acquisitions provided multitudinous opportunities for many aggressive and adventuresome pioneers. Yet at the same time, the westward movement that resulted created special problems of frontier defense against the militant natives.

While the permanent fortifications for the defense of the sea frontiers were rising along the Atlantic and Gulf coasts, the army was also engaged with the defense of land frontiers. Inland, however, the problem of defense was considerably different, since the type of hostilities differed. Indians conducted a primitive form of warfare on a frontier that was constantly changing with the settlement of new territory. Among the most challenging problems encountered by military engineers was the development of works appropriate for the nature of the hostilities of the Indian tribes in the various areas and the adaptation of the form of their architecture to the climate and materials of each particular region. There was no type of fortification, system of building, or style of archi-

tecture that was universally adaptable to the diverse conditions encountered. Only time and experience could reveal the best solutions to local problems.

LATE-EIGHTEENTH-CENTURY FRONTIER DEFENSE

When the settlers moved inland from the Atlantic, Indians continued to terrorize the frontier just as they had many of the earliest settlements along the coast. Early settlers in New England built their own log garrison houses, blockhouses, and stockades for defense. In the westward movement during and immediately after the Revolution, survival in the new settlements of Kentucky, Ohio, and western Pennsylvania required structures for defense against Indians prodded by the British, and again the settlers for the most part fended for themselves by constructing communal stockaded works such as Boonesborough, Kentucky (1776), and Farmers' Castle, Ohio (1791). Only after the war was it possible to raise military forces to undertake the construction of forts to assist in land frontier defense.

Military defensive structures in the wilderness were comparable to the civil structures that had been erected by the settlers. Located in the same regions, their function was essentially the same: to provide security against Indian raids. For this purpose, enclosures of stockades were quite adequate; earth or masonry works were not needed.

Although primitive types of enclosure were commonly used, early plans were artfully conceived. A stockade developed on a five-bastioned trace was used to enclose Fort Harmar, Ohio (1785), which was built to defend the western frontier (fig. 74). As was common in bastioned works, the buildings within the enceinte were neatly located around a parade and were set parallel to the curtains. Like Harmar, Fort Fayette, Pennsylvania (1792), also had a bastioned enclosure (fig. 75). Constructed under the direction of Isaac Craig near the site of Fort Pitt,[1] Fayette was a square, four-bastioned fort with an enceinte of twelve-foot-high stockading. In three bastions there were two-story blockhouses with cannons on the second floor which could fire over the stockade, and in the fourth bastion was a bombproof magazine.

To protect and control new settlement, the military pushed westward with the construction of other forts. In western Ohio was Fort Washington (1789), a log structure with blockhouses instead of bastions on the corners of the stockade (fig. 76). North, and yet farther west, were Fort Recovery, Ohio (1793), Fort Wayne, Indiana (1794), and Fort Defiance, Ohio (1794), all established by Anthony Wayne. Recovery and Defiance were stockaded forts with blockhouses at each corner, while Fort Wayne was a stockaded work with blockhouses on diagonal corners.

While the plans for enclosures that were developed on bastioned traces provided effective configurations for the defense of the stockade, these simpler arrangements, which required less perimeter to enclose the same amount of area, were more common. In primitive defense, the simple, rectangular stockade could be adequately enfiladed on all four sides with muskets from blockhouses at diagonal corners, if the fort was small, or from blockhouses at all corners if the stockade was lengthy on each side—arrangements also commonly used in civil frontier defense.

Fort Dearborn, Illinois (1803, 1816), with its stockade and diagonal blockhouses designed by Captain John Whistler,[2] conformed to the diagonal blockhouse arrangement (fig. 77). Since the corner structures were the keys to the defense of the fort, they were naturally the strongest works and were built with timbers either hewn or sawn into rectangular cross section. While the plan for the body of Fort Dearborn was common, the second stockade located outside to create a second line of defense was unusual on the frontier.

THE SOUTHEASTERN FRONTIER

The simple rectangular enclosure and blockhouses were also common in Florida during the conflicts with the Seminole Indians from 1816 to 1818 and 1835 to 1842. Located on an eminence, Fort Mitchell, Alabama (1813, 1828), was "a square formed by pickets with a blockhouse at two diagonal corners." Fort Harllee, Florida (1836), was likewise described as consisting of "pickets with two blockhomes at diagonal corners, after the fashion of Florida forts."[3] As in the past, this arrangement allowed an efficient defense, with the stockade providing protection against surprise attack and with each blockhouse flanking two walls.

Apparently, in the primitive areas of Florida all stockades were similar. According to a contemporary description: "Pickets are made by splitting pine logs about eighteen feet in length into two parts, and driving them upright and firmly into the ground close together, with the flat side inwards; these are braced together by a strip of board nailed on the inside. The tops are sharpened, and holes are cut seven or

Land Frontier Forts

74. Fort Harmar, Ohio (1785). Captain John Doughty, engineer. Painting by Wyllys Hall. *Marietta College, Marietta, Ohio.*

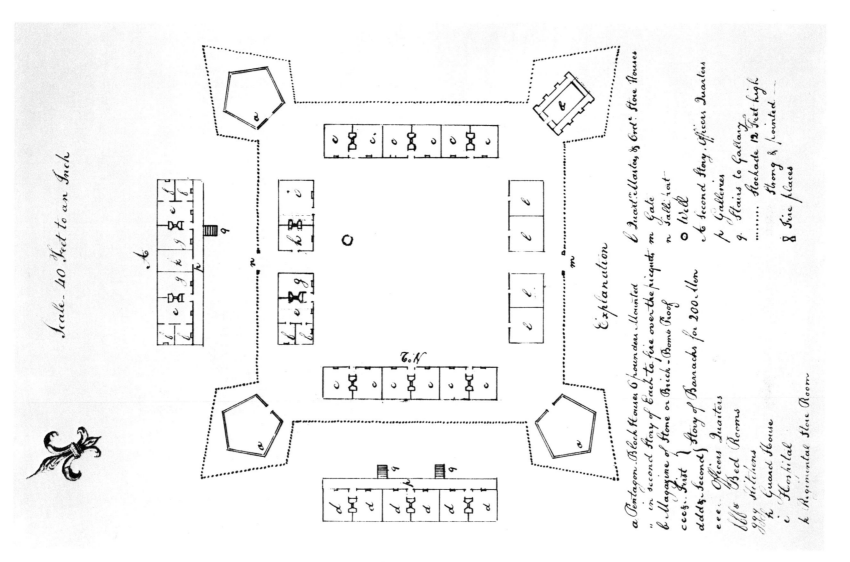

75. Fort Fayette, Pennsylvania (1792). Isaac Craig, engineer. *William L. Clements Library, Ann Arbor, Mich.*

76. Fort Washington, Ohio (1789). Lithograph after a painting by H. W. Kemper. *National Archives, Washington, D.C.*

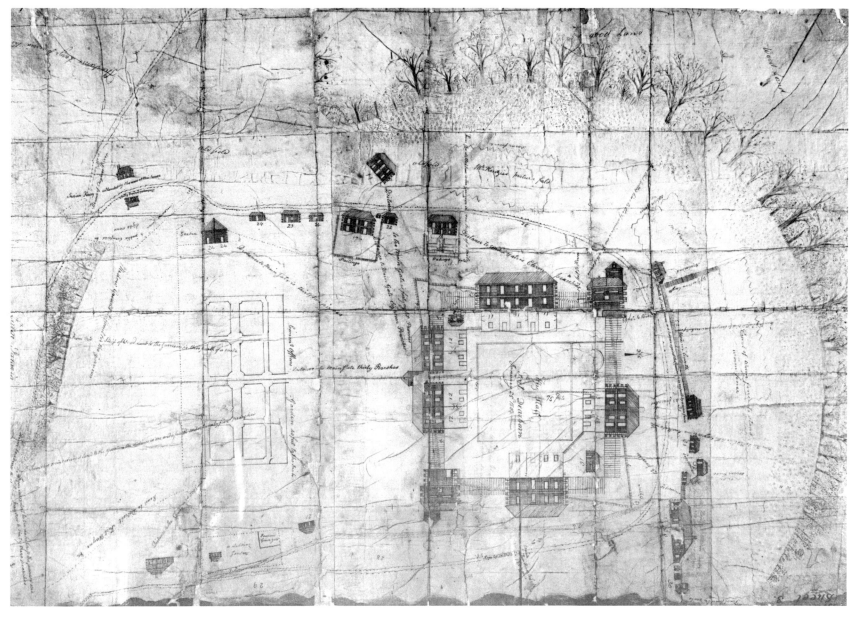

77. Fort Dearborn, Illinois (1803, 1816). Captain Whistler, engineer. Sketch (1808) by engineer. *National Archives, Washington, D.C.*

eight feet from the ground for the fire arms. A range of benches extends around the work about three feet high, from which the fire is delivered."[4] The benches, of course, allowed the loopholes to be elevated so they could not be used by an enemy from the outside.

While similar techniques were used in most of the forts constructed during the Seminole War, there were obviously contrasts in the magnitude of the works and the architectural forms used under various conditions. Fort Mellon on Lake Monroe (ca. 1836) was simply a stockaded enclosure with no other defensive works. Like Fort Volusia (ca. 1836) to the north, it was abandoned because of an unhealthy location and was subsequently burned by Indians.[5] Fort Jupiter, Florida (1838), intended as a supply depot, likewise consisted only of a stockade enclosure.

Other works appeared evidently without stockades. Fort Peyton, Florida (ca. 1836), "consisted of four log houses built in a hollow square; two occupied by the troops; one by the officers, the fourth used as a hospital and commissary store." Fort Pierce (1838)—doubtfully deserving the appellation of "fort"—was a blockhouse "pretty much like all other block-houses in Florida, except that this one was built of palmetto logs."[6] All these works were simple in form and primitive in construction. Although the forts of the Seminole wars had in common the basic purpose of defense against Indian raids, a variety of other conditions—locations, duration of use, number of defenders, and so on—resulted in some diversity of form.

These forts were expedient and temporary affairs. Expediency was mandatory because of the exigency of immediate physical protection. The need for efficiency was given more impetus by the relatively small number of officers and men sent to enforce the orders of the administration. Erected as the need arose, the period of use of most of these forts was ex-pected to be short; permanency was not an important consideration.

THE MIDWESTERN FRONTIER

Other Indian problems developed after the Seminoles were eventually suppressed and many were moved. During the first part of the eighteenth century entire tribes from the East were relocated on lands west of the Mississippi. The relocated tribes and the natives who already occupied the land further extended the obligations of the army. These Indians, along with settlers, created a threefold purpose for military defense along the frontier that was then the West but is now the Midwest: to maintain peace among the Indians, to protect the immigrants, and to guarantee the treaty protection of peaceful weak tribes against the strong hostile nations. To fulfill these objectives, a system of military posts was required along the frontier.

By 1830 a chain of forts along a north-south line had been established. The eastern border of Texas defined the southern end of the line, which then approximately followed the western boundaries of the present states of Arkansas and Missouri, the Missouri River for a short distance, and thence a line across to Lake Superior. Scattered along this fringe, with the intent of defending the entire frontier, was a cordon of posts which were often called the advanced or exterior line of defense.[7] Key among these were Fort Snelling (1819), Fort Leavenworth (1827), Fort Smith (1817, 1838), and Fort Towson (1824), all located in widely separated areas of the wilderness which later became, respectively, the states of Minnesota, Kansas, Arkansas, and Oklahoma.

To preserve peace, forts near the advanced line of defense were at first well within the country assigned to the Indians. While the general locations were determined by the character of the frontier, specific sites were selected with regard to several fundamental requirements: communication, defensive advantages, and availability of good natural building materials. Locations adjacent to navigable rivers, eminences which commanded all approaches, and convenient sources of timber, stone, or clay for brick were considered essential. In addition, attention was given to the availability of potable water, fuel—either timber or coal—and hay for the livestock.

Although it is taken for granted now, the defense of the western Indian frontier contained several uncertainties for military architects in regard to construction and fortification. Unlike construction in Florida, where it was known that most forts needed to be only temporary affairs, it seemed that durable architecture was mandatory in the West, since the western frontier at that time appeared to be somewhat stable. But the type of war that would be waged by the Indians could not be exactly anticipated; hence, the nature of appropriate defensive works was indefinite. Past history of Indian conflicts—such as Pontiac's War—perhaps provided some clues but no certain answers.

In the design of the forts of the advanced line there were various interpretations and reactions to these problems of construction and defense. Fort Snelling was located on a point of land defined by an almost perpendicular cliff at the confluence of the Minnesota and Mississippi rivers (fig. 78). Established by Colonel Josiah Snelling, it consisted of a series of buildings arranged around a parade and enclosed by a nine-foot-high wall which followed a diamondlike trace. Flanking the walls at acute angles of the enclosure were round towers; at the obtuse angles were hexagonal towers. Each was provided with tall, narrow rifle ports directed both to the interior and to the exterior of the enclosure.

The walls, towers, and buildings, all naturally and beautifully integrated into the site, were durably constructed of stone, making Snelling a permanent fortification. Over a long period of occupation, this type of construction proved to be most economical. Yet in 1829 Major General Edmund P. Gaines, commanding officer of the Western Department, reported that the plan was defective and the construction excessively elaborate. Buildings, he wrote, were "too large, too numerous, and extending over a space entirely too great, enclosing a uselessly large parade." He further observed that much credit was due "for the *immense labor and excellent workmanship exhibited in the construction . . . of barracks and storehouses,*"[8] but that this construction was effected at too much expense of the discipline of the troops who worked on them. Gaines believed that the soldiers should have spent more time drilling and less time exercising their building skills.

Fort Leavenworth, established and presumably planned by Colonel Henry Leavenworth (1783–1834), was likewise conceived on a large scale with durable construction (fig. 79). Among the first permanent works undertaken on the hill selected as a site may have been a stone wall with rifle ports, the intended extent of which is not now certain. Like several other forts built during the same period, Leavenworth was originally called a "cantonment" because the first quarters were temporary. However, the first works were rapidly replaced by substantial structures, and by 1832, when the post was designated a fort, there were brick and wood-framed officers' quarters, barracks, and a hospital, all arranged in orderly fashion.

Of all the forts designed for Indian country, the concept for the second Fort Smith was among the most extensive (figs. 80, 81). Replacing the first group of buildings, which were in the form of a hollow square with blockhouses on diagonal corners, the plan for the new fort enclosed a large area on an elevated site overlooking the Arkansas River. A five-bastioned enceinte was proposed by the designing engineer as if it were expected that a large force of Indians would besiege the place. How-

78. Fort Snelling, Minnesota (1819). Established by Colonel Josiah Snelling. Painting (1844) by J. C. Wild. *Minnesota Historical Society, Saint Paul.*

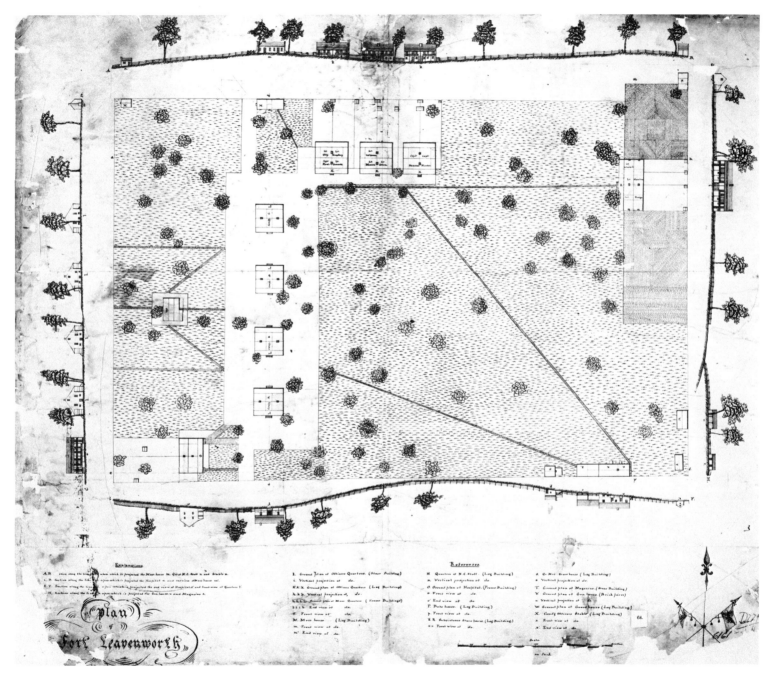

79. Fort Leavenworth, Kansas (1827). Established by Colonel Henry Leavenworth. Plan (1834). *National Archives, Washington, D.C.*

80. Fort Smith, Arkansas (1838). Plan. *National Archives, Washington, D.C.*

81. Fort Smith, Arkansas (1838). Drawing (1853) by Möllhausen. *Oklahoma Historical Society, Oklahoma City.*

82. Fort Gaines (Fort Ripley), Minnesota (1849). Painting (ca. 1860) by Franz Hölzlhuber. *Glenbow-Alberta Institute, Calgary, Alberta.*

ever, the fortifications were never completed. Danger subsided by 1841, and the post was transformed into a supply depot. Only one part of the wall had been built; a partially completed bastion was transformed into a commissary.

Experience revealed that the fortifications at Fort Snelling and those projected for Fort Smith were considerably more elaborate than any required against the Indians. The Indians occupying the western country were not prepared to besiege a fortified place; they wisely preferred to do most of their fighting against small, isolated parties. As the frontier developed, this fact was recognized, and measures for active defense of forts were simplified. At Fort Gaines (Fort Ripley), Minnesota, for example, blockhouses were retained, but complete enclosure was eliminated (fig. 82). Later, most western forts became simply open complexes of buildings with no fortifications, although they were known by the appelation "fort."

REGIONAL DEFENSE

In Texas a north-south chain of unfortified posts was established near mid-century to protect settlements west of the Trinity River from hostile natives. Begun in 1849 were Forts Worth, Croghan, Gates, Graham, and Inge, geographically located in nearly a straight line running northeast to southwest. It was intended that the Indians remain north and west of this line. The general positions of these forts were established after evaluating the nature of the frontier to be defended, while specific sites were selected which were near water yet were well drained. However, these forts were to be temporary, for the line of defense would move westward as the frontier advanced.

The plan of Fort Worth, the northernmost of this chain, was characteristic of these unfortified posts as well as of many other forts which were subsequently constructed (fig. 83). Established by Brevet Major Ripley Arnold on a bluff overlooking the Trinity River, the buildings were arranged in an orderly pattern around a broad, rectangular parade oriented at an angle with the cardinal points. Typical military hierarchy was observed in the position of buildings. Officers' quarters occupied the southeastern side of the quadrangle, with barracks on the opposite side. On the northeast, convenient to both officers and enlisted men, were the stables. The fourth side of the parade was partially closed by the adjutant's office, from which were transmitted the commanding officer's orders, and by the hospital. This location for the hospital was somewhat atypical, since it was not, in other posts, usually included in the formal grouping. Service buildings (kitchens, laundresses' quarters, and mess hall) were conveniently located behind the structures they served.

Characteristic of the other Texas forts commenced in 1849, the post on the Trinity was destined for a short existence of five years; hence, the buildings were transient constructions built of materials readily at hand. In fact, orders pertaining to the establishment of the post had specified that "No permanent construction of quarters &c. will be made, until... sanctioned by the Superior authority."[9] The three shelters housing officers and the commissaries located on the south corner were of logs; one of the barracks, the mess hall, and the stables, most of which were built by the troops, were described as palisade works—a technique of construction which resembled the French *poteaux en terre* earlier used in colonial architecture in Canada and Louisiana. Similar to stockades in principle, the walls were of logs which were uniform in length, planted tightly together on end in the ground, and topped by a log plate. Typically, the interstices were then filled with wood scraps and mud. Several other buildings were

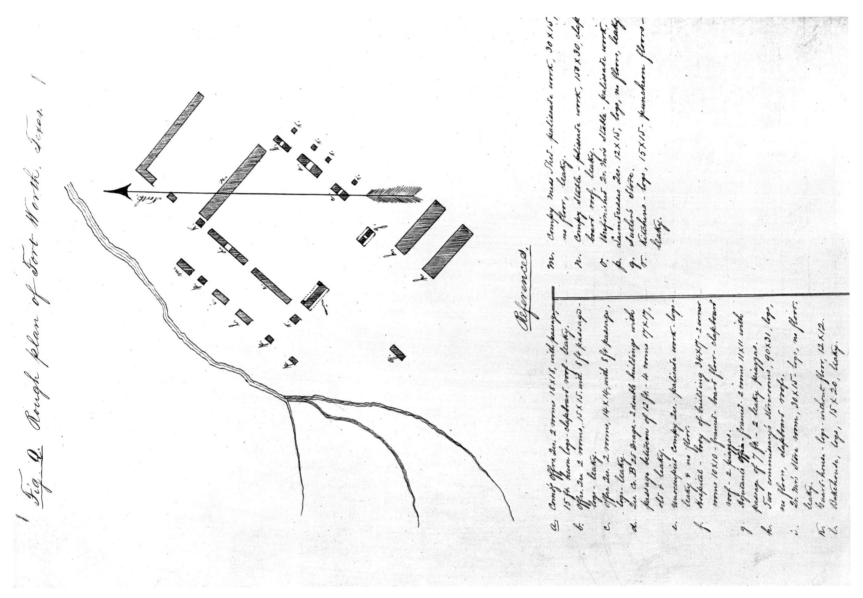

83. Fort Worth, Texas (1849). Established by Brevet Major Ripley Arnold. Site plan. *National Archives, Washington, D.C.*

"frame . . . ceiled with inch boards and roofed with shingles."[10]

Temporary construction also characterized military buildings near the Mexican border. Employing techniques that were basically similar to those used at Fort Worth, several of the shelters at Fort McIntosh, Texas (1849), were structures called *jacales*, with a type of construction that had been used by both Spaniards and Mexicans. One building, used for officers' quarters, was described as "twelve by sixteen feet, made of mezquite poles placed perpendicularly in the ground, dirt floor, shingle roof." Two others, one used by officers and the other used as a carpenters' shop, were " 'jacal' . . . covered with paulins."[11]

But even with temporary construction, in the most primitive areas a strong sense of order was always expressed. As the frontier moved westward in Texas, concepts of planning remained the same as those of Fort Worth. Forts Mason, Belknap, McKavett, and a post on the Brazos, Fort Phanton Hill, a second chain of posts established between 1851 and 1852, were similarly planned in formal arrangement. However, the sites for the second cordon differed from those of the first chain in climate, terrain, and geological formations. Roughly parallel with the first line, the second occupied the eastern edge of a more arid region. Limestone was plentiful, so it became the major building material used in the erection of post buildings.

In this second chain of forts, although stone and clay suitable for brick were at hand, materials for framing, finishing, and roofing had to be transported to the site from distant locations. For Fort Belknap, Texas (1851), for example, squared timber was hauled seven to twenty-three miles, while oak shingles were carried forty-five miles (fig. 84).[12] The story of the building of Fort McKavett (1852), established by Major Pitcairn Morrison, was similar. It was located on top of a hill in an area devoid of trees suitable for lumber, and necessary wood materials for the post on the edge of the Edwards Plateau were thirty miles away.[13]

In the Texas posts the buildings exhibiting the most durable structure, as attested by present-day ruins, generally were the powder magazines. At the post on the Clear Fork of the Brazos (Fort Phantom Hill) (1852), established by Major John Joseph Abercrombie (fig. 85), the magazine was roofed with a stone vault supported by thick walls and buttresses. A correspondingly finer workmanship also characterized this and other similar structures.

The stone forts of central Texas were spasmodically occupied and abandoned for relatively short periods of time, reflecting the unstable condition of the frontier. The post on the Clear Fork of the Brazos was first abandoned only three years after it was established; Fort Belknap was abandoned in 1859 because of a shortage of water, but it was reoccupied in the sixties; Fort McKavett was likewise abandoned and reoccupied several times before its permanent evacuation in 1883.

MILITARY ARCHITECTURE OF THE ARID AND SEMIARID REGIONS

Different climatic and geographical features of other regions provided different conditions for location and construction. In areas where rain was scarce and fighting primitive, the locations of springs or prospective wells were of primary importance. Positions flanked by hills, bluffs, or mountains were therefore often selected because of the proximity of water at their bases. Thus, forts in the arid Southwest were often situated at the entrances to canyons.

Described in 1853 as "the most beautiful and interesting post as a whole in New Mexico,"[14] Fort Defiance, Arizona (1851), authorized for the purpose of controlling the warlike Navajo Indians, was located at the mouth of the steep-walled

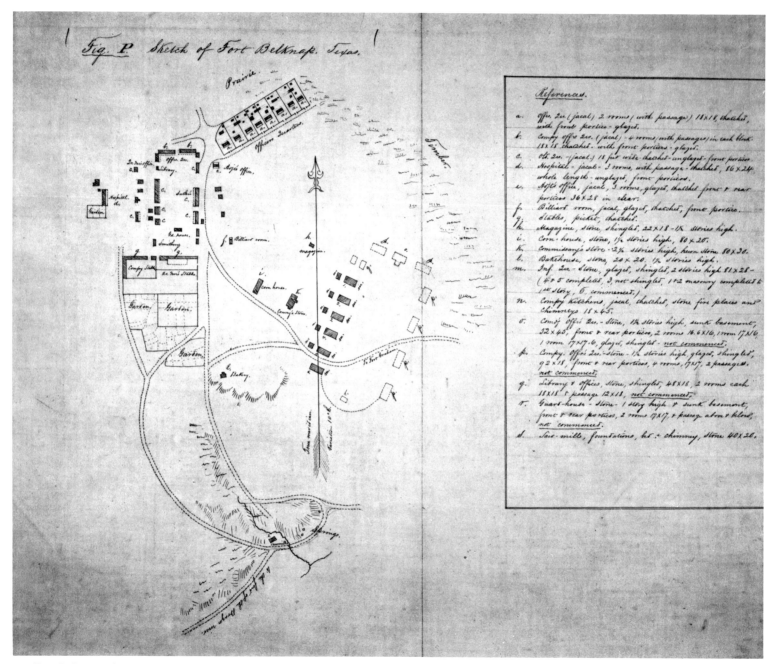

84. Fort Belknap, Texas (1851). Established by Lieutenant Colonel William C. Belknap. *National Archives, Washington, D.C.*

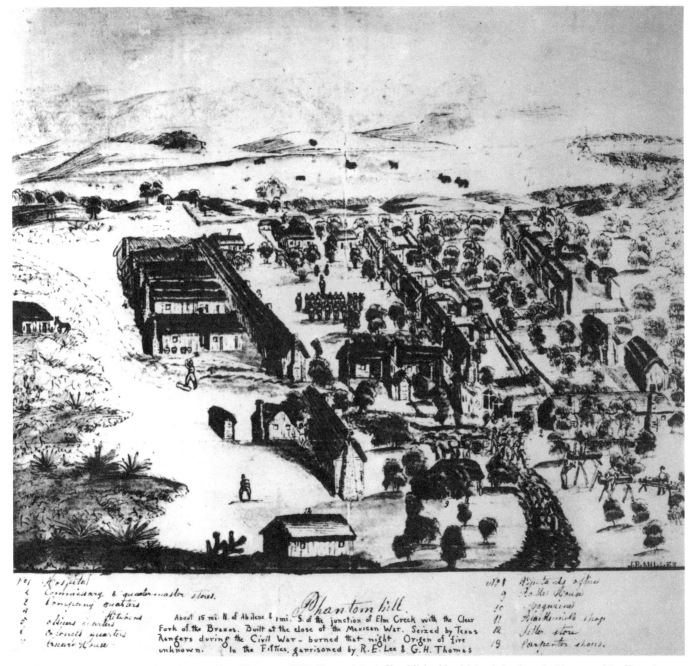

No 1   Hospital
2   Commissary & quartermaster stores.
3   Company quarters
4   Officers quarters   Kitchens
5   Colonels quarters
    Guard House

Phantom hill.

About 15 mi. N. of Abilene & 1 mi. S. of the junction of Elm Creek with the Clear
Fork of the Brazos. Built at the close of the Mexican War. Seized by Texas
Rangers during the Civil War – burned that night. Origen of fire
unknown.   In the Fifties, garrisoned by R.E. Lee & G.H. Thomas

No 1   Adjutants office
9   Sutlers House
10   Magazine
11   Blacksmiths shop
12   Sutler store
13   Carpenter shops.

85. Post on the Clear Fork of the Brazos (Fort Phantom Hill), Texas (1852). Established by Major John Joseph Abercrombie. Drawing by J. R. Miller. *Barker Library, Austin, Tex.*

86. Fort Defiance, Arizona (1851). Established by Colonel Edwin V. Sumner. Watercolor by Seth Eastman. *Boston Museum of Fine Arts.*

Cañon Bonito near water supplied by a spring-fed stream (fig. 86). In this canyon and others nearby there was also good forage, and on or near this isolated site in the wilderness soldiers found earth, timber, and stone with which to build. Originally under the supervision of Major Electus Backus and subsequently under Brevet Lieutenant Colonel Joseph H. Eaton,[15] many of the structures at Defiance were constructed from earth and wood. The walls of most consisted simply of pine logs laid up horizontally. Only a few were adobe, while one building for officers was of stone.[16]

The form of the buildings was developed in response to the climate. Since this was an arid region, there was little need for pitched roofs which would shed rain or snow. Therefore, undressed logs simply spanned from wall to wall, supporting a flat dirt roof. Earth was a good insulator against both cold and heat. Although it would naturally leak during heavy rains, it could absorb light amounts of moisture.

When the soldiers moved into unfamiliar but previously inhabited areas of the Southwest, they by necessity adopted building techniques that were indigenous to the area if those techniques were applicable to their needs. These building methods had proved to be the most advantageous for their particular area over a period of many decades. Moreover, when local civilian workers were hired, they then worked with a familiar type of construction.

The primitive, circular hogans of the Navajos in northwestern New Mexico and northeastern Arizona, in the vicinity of Fort Defiance, suggested little which was appropriate for military architecture, but the Spaniards and Mexicans had been using methods that were highly adaptable. Thus, when Fort Craig, New Mexico (1854), was established along the Rio Grande, the Spanish technique of building with adobe was employed (fig. 87). A large, rectangular adobe enceinte which was contiguous at several places with walls of interior buildings enclosed the fort as defense against the dreaded Apaches, who roamed at large. These thick curtains were flanked by irregular bastions located at each corner. Inside the enclosure, further Spanish-Mexican influence was seen in the plans of the barracks, "two in number, built of adobe in the form of a hollow square, each enclosing a plazita."[17] One was occupied by white troops and the other by black soldiers.

Of the posts in the Southwest, the assimilation of the Spanish style of building was also typified by the structures of Fort Sumner, New Mexico (figs. 88, 89). Designated as a nucleus for civilizing the Navajo and Apache tribes, Fort Sumner was begun in 1862 on a site selected on the east side of the Pecos River.

Although the orderly arrangement of the buildings and the zoning of the site into distinct functional groups of buildings respected American military tradition, the integration of courts into the scheme suggests its adaption from early Spanish architecture. With one exception on the north, enlisted men's quarters were oriented around enclosed courts; those of officers were positioned to create courts open in the direction of the parade. All of these units were connected to create a parade that was accessible only from the four corners. The use of courts also prevailed outside the dwellings in the hospital, Indian commissary, stables, and quartermaster structures.

Further examples of forts reflecting endemic influences were seen in other areas of the Southwest. At Fort Union, New Mexico (1851), new buildings were begun in 1863 to replace earlier ephemeral structures (figs. 90, 91). Located east of the original works and near an earlier bastioned earthwork,[18] these buildings were of adobe bricks manufactured from the soil of a valley north of the fort.[19] As in other remotely located forts with no timber at hand, lumber as well as hardware and other finishing materials had to be hauled to the site.

A four-company post and regional supply depot, Fort Union was an elaborate work with many buildings, all clearly located with respect to military tradition. The orderly positioning of buildings around the parade was characteristic of most earlier posts and recalled the formality of an ancient Roman military camp. The residence of the commanding officer toward the northwest corner of the parade terminated a vista created by a lane between the opposite barracks which gave access to the chapel, stables, and prison. Only quarters and offices for ranking personnel were located on the southwest.

Fort Union had been strategically located at the junction of the Mountain Branch and the Cimarron Cutoff of the Santa Fe Trail. To protect emigrants traversing the Lower Road through southern Texas between San Antonio and Fort Bliss (1848), Forts Davis (1854), Lancaster (1855), and Quitman (1858) were established. As in other areas, military designers adjusted their building techniques to the conditions.

At Fort Davis characteristic patterns were followed. Many of the first buildings were temporary, some with walls of boards and battens or of poles planted vertically into the ground and roofs of thatching. Then, after a period of vacancy during the Civil War, a new post was established on a site near the first.

In 1867 the second Fort Davis (fig. 92) was planned by Lieutenant Colonel Wesley Merritt (1834–1910), who was later to be appointed to the superintendency of West Point. Merritt made large plans: locating his post on a flat plain just outside the entrance to rugged Limpia Canyon, his design called for a formal arrangement of many buildings around a monumental parade nearly four hundred feet wide and eleven hundred feet long. Officers' quarters on one long side faced barracks on the other. A chapel and the headquarters building were emphasized by their locations at one end.

While the formal planning of the post sharply contrasted with the rugged land forms, the materials harmonized. Some of the first buildings were constructed substantially, with the help of civilian craftsmen, from stone that was quarried locally. However, searching for less time-consuming methods appropriate for the region, the builders adopted the adobe technique, which resulted in a large percentage of adobe-brick structures on stone foundations. Of the nineteen hipped-roofed officers' quarters built by the end of the 1860s, four had walls of limestone while the others were adobe. Four barracks were also built with adobe and were plastered inside and out.[20]

Adobe construction in military architecture became widespread throughout the Southwest. Fort Davis's sister forts—Lancaster and Quitman—had buildings with thick walls of sun-dried bricks. However, in those structures where security was important, such as guardhouses, stone continued in use.

THE FAR SOUTHWEST

Military posts established in the arid areas of the Far Southwest throughout the 1850s, 1860s, and 1870s also had buildings that were predominantly of adobe. In 1875 it was reported that all the buildings at Fort Yuma (1850)—called "the hottest fort in the country"—had walls "three feet thick, . . . composed of double adobe walls, with a space between."[21] Thick walls separated by an airspace provided good insulation against the heat. The buildings at Forts West (1863) and Bayard (1866) in southwestern New Mexico were likewise mostly adobe, as were those at Forts Lowell (1862), McDowell

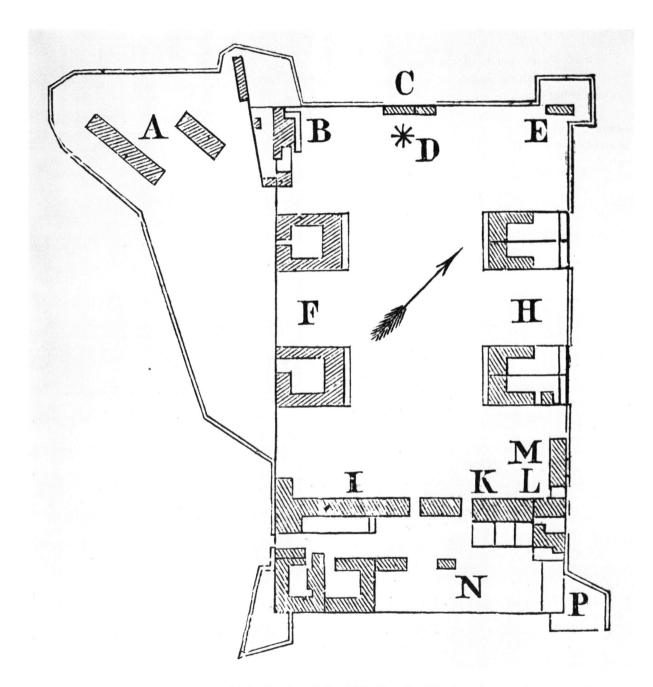

87. Fort Craig, New Mexico (1854). Established by Captain Daniel T. Chandler. Plan from Surgeon General's Office, *Circular No. 4: A Report on Barracks and Hospitals with Descriptions of Military Posts.*

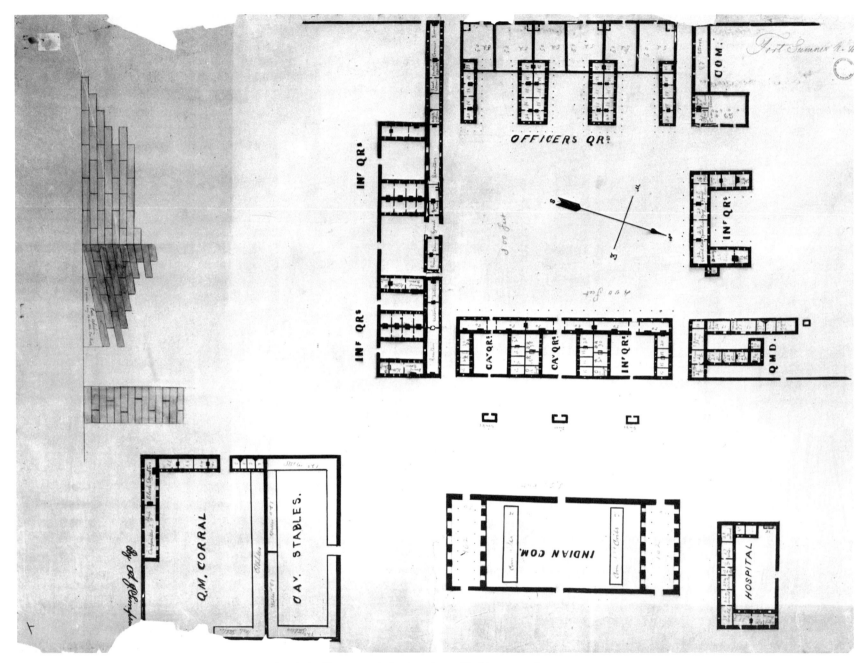

88. Fort Sumner, New Mexico (1862). Plan drawn by A. J. Simpson. *National Archives, Washington, D.C.*

FORT SUMNER, N.M.

VIEW FROM EAST SIDE.

Fellner

89. Fort Sumner, New Mexico (1862). Drawing by Fellner. *National Archives, Washington, D.C.*

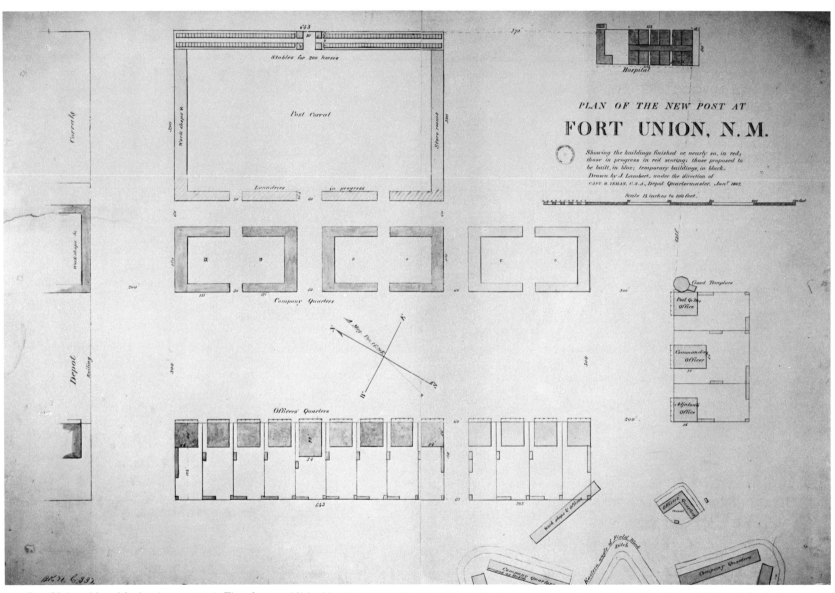

90. Fort Union, New Mexico (1851, 1863). First fort established by Lieutenant Colonel Edwin V. Sumner. Second fort planned by General Edward R. S. Canby. Plan (1867) drawn by J. Lambert. *National Archives, Washington, D.C.*

91. Fort Union, New Mexico (1851, 1863). Drawing by Joseph Heger. *Arizona Pioneer's Historical Society, Tucson.*

92. Fort Davis, Texas (1854, 1867). First fort established by Lieutenant Colonel Wesley Merritt. Drawing (1867) by Leon Grousset. *National Archives, Washington, D.C.*

(1865), and Thomas (1876) in southern Arizona and Fort Churchill (1860) in western Nevada.

Near the Pacific Coast, natural conditions were unlike those of southeastern California, Arizona, New Mexico, and western Texas. Since the coastal Indians were mostly peaceful, little stress was placed on active defense; thus, virtually no fortifications were required. Near the coast, forts were therefore generally small and built of materials found near the sites.

As in other regions, regardless of the nature of the terrain formal arrangements of buildings were maintained.

At Fort Miller, California (1851), a post established by First Lieutenant Treadwell Moore to protect miners in the San Joaquin Valley, buildings were placed in the conventional arrangement (fig. 93). The parade, a small, rectangular area measuring two hundred feet by three hundred feet, was enclosed by a low, stone-capped adobe wall which must have

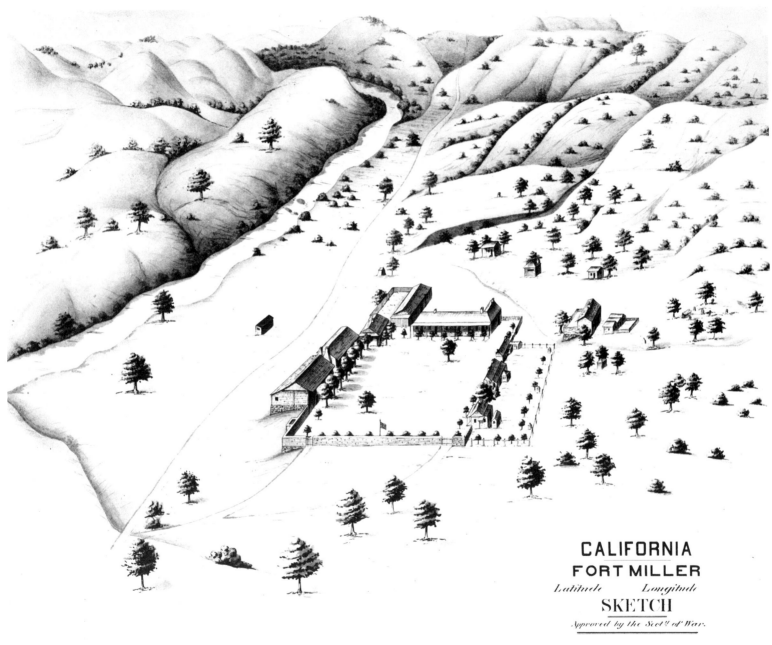

CALIFORNIA
FORT MILLER

Latitude   Longitude

SKETCH

Approved by the Sect.^y of War.

Copied from drawing on file in Q.M.G.O. by C. Jacobsen

93. Fort Miller, California (1851). Established by First Lieutenant Treadwell Moore. *National Archives, Washington, D.C.*

proved to be more of a barrier to cattle than to Indians. Adobe was also used in the construction of part of the buildings, among them the officers' quarters. Other functions were housed in log or canvas shelters.

Disregarding size, a main difference between the forts of California and those of the dry areas to the east was in the roof forms. To shed heavier rainfalls, buildings near the coast had pitched roofs finished with shingles, instead of flat or low-pitched surfaces of dirt or tin. This was true not only for adobe walls, which had to be protected from moisture with roof overhangs and plaster finishes, but also for log buildings. Buildings at Fort Miller, except for the makeshift affairs, had gable or shed roofs covered with shingles.

Buildings of other forts established in the early 1850s in the region of northern California where there was abundant timber were of wood. At Fort Jones (1852), "the officers and soldiers quarters, and store rooms, and hospital, and stable, were of logs, and erected by the men. Of course quite indifferent, but such as other people enjoy and sufficient for the present."[22] Fort Lane, Oregon (1852)—also a part of the Pacific chain—was likewise composed of log buildings. Apparently slightly more refined in construction, the buildings at Fort Humboldt, California (1853), were "frame and boards."[23]

Forts on the Plains

Settlers destined for the Oregon Country to the north traversed the lengthy Oregon Trail. Across dusty plains and through rugged mountain passes they followed a natural highway determined by the Platte, North Platte, Snake, and Columbia rivers. As in the earlier history of the eastern section of the country, routes of communication determined the locations of forts. Established for the defense of the trail against Indian hostilities were Fort Kearny, Nebraska (1848), Fort

Laramie, Wyoming (1849), Fort Bridger, Wyoming (1842, 1857), Fort Hall, Idaho (ca. 1855), and Fort Vancouver, Washington (1824–48). Of these, all but Fort Kearny were on or near sites where civilian trading posts had previously been established.

The military forts on the wind-swept prairies were similar in layout to many of those in the Southwest but had a different architectural character as a result of their natural setting and construction. At Fort Kearny the buildings were clustered around a parade, thus embodying the typical desire to create order in the wilderness (fig. 94). Located to the southeast, Fort Riley, Kansas (1853), which eventually became a post of many buildings in the Romanesque Revival style, was similar in concept (fig. 95). The long horizontal lines of roof overhangs and the horizontal rhythm of openings created a scene of repose between the buildings and the expansive plains. But a welcome visual relief was provided by the pitch of the gables—all of this a spontaneous expression of function and natural conditions in a rude wilderness. A similar quality was found at Fort Laramie, except that in the roof angles there was some repetition of the hill forms around the site (fig. 96).

Fort Laramie, one of the most active posts in the West, was located on a site adjacent to the Laramie River after a reconnaissance of the surounding area proved that there was no location with greater potential.[24] While conforming to military tradition, the plan of the fort was thoughtfully adjusted to topography (fig. 97). The site was a narrow bluff created by the erosive river current. Three flats at different elevations were within the area encompassed by the fort, thus providing a natural, orderly zoning of functions: on the highest level was the cemetery; the intermediate area was given to buildings housing regular military functions; the bottom level was occupied by service facilities.

Early plans indicate that fortifications were intended for

94. Fort Kearny, Nebraska (1848). Established by Lieutenant Colonel Ludwell E. Powell. View from northeast (1870); painting by Anton Schonborn. *Amon Carter Museum of Western Art, Fort Worth, Tex.*

Fort Laramie. An 1851 drawing shows a proposed rectangular enclosure with blockhouses on three corners. In 1863 no defensive works were indicated,[25] but the plan of the works as they existed in 1867 suggests a trace for fortifications on the northeast side only. It was found that the complete defensive enclosures that were often urgently needed for the protection of small numbers in civilian forts were unnecessary at Fort Laramie, where many defenders were present. In fact, friendly Indians often wandered freely among the buildings.

In the decade following the establishment of the fort on the Laramie River, with Indians threatening, other posts were built to protect communications in the present state of Wyo-

ming. They included Fort Fetterman (1867) and Fort Fred Steele (1868). Like most of the other prairie forts near the Oregon Trail, these posts were simply open groupings of buildings (figs. 98, 99). Fort Fetterman and probably Fort Fred Steele were "ordered to be built of such materials as the region afforded."[26] In the same region, and if the nature of defensive functions did not change, architecture remained consistent in form and arrangement.

Although the open plan and characteristic formality were retained, the conventional rectangular arrangement of buildings on the site was abandoned at Fort D. A. Russell, Wyoming (1867), another post situated on the broad prairie (fig.

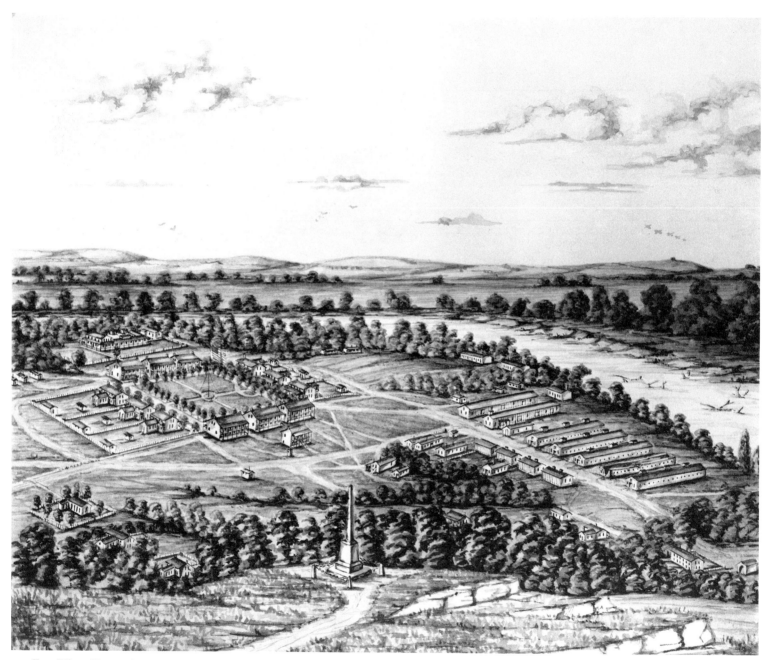

95. Fort Riley, Kansas (1853). Established by Captain Charles S. Lovell. Watercolor (1873) by unknown painter. *New York Historical Society, New York City.*

96. Fort Laramie, Wyoming (1849). Established by Colonel David E. Twiggs. View from the east (1870); painting by Anton Schonborn. *Amon Carter Museum of Western Art, Fort Worth, Tex.*

100). According to a report by Surgeon C. H. Alden, the different site plan developed from considerations for environmental conditions and appearance:

The barracks do not directly face the parade, but are arranged *en echelon*, by which means light and air have free access to all sides of the buildings. . . .

The general plan of the post was made by Bvt. Brig. Gen. J. D. Stevenson . . . , with suggestions by Surgeon C. H. Alden. . . . The diamond form of the parade was adopted not only for the sake of appearance, but to avoid the inconvenience of the very large inclosed space, which would have resulted from the ordinary rectangular or square enclosure.[27]

The buildings reflected the climate and the times. Barracks were framed with wood and had walls that were finished on the outside with rough boards and battens, while the roofs were shingled. For insulation, between the wall timbers there was "a lining of adobes, placed on edge . . . to the level of the eves all around."[28] Adobe-lined frame walls also appeared in buildings at other prairie forts, among them barracks and officers' quarters at Fort Laramie. At D. A. Russell, at least two buildings were prefabricated. The commissary and the quartermaster's storehouse, about twenty-five feet by one hundred feet, were built in a style "known as sectional, they having been brought up from Omaha in parts, and put together" on the site.[29]

On the plains north of the Oregon Trail, other forts were established in the fifties, sixties, and seventies to control the warlike Indians of that region. These forts also functioned to

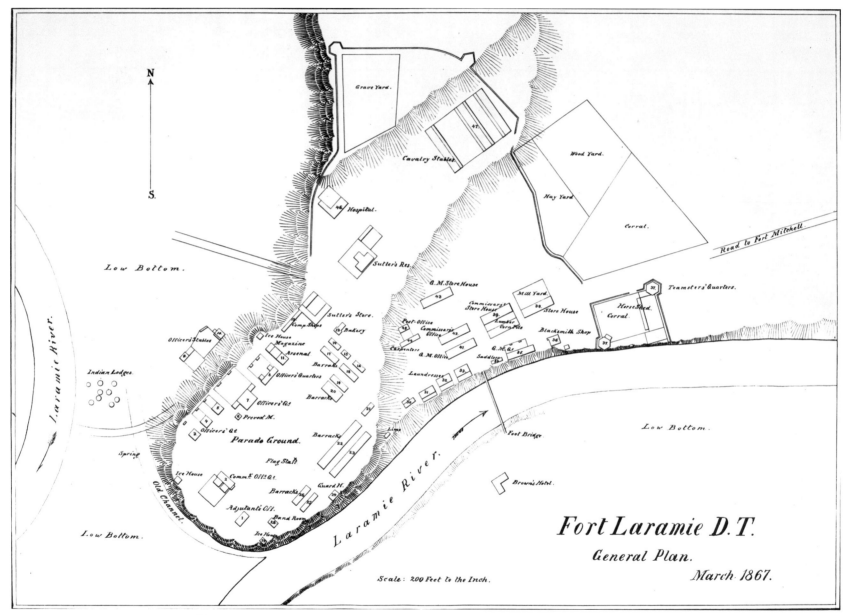

Fort Laramie D.T.

General Plan.

March 1867.

Scale: 200 Feet to the Inch.

97. Fort Laramie, Wyoming (1849). Plan (1867). *National Archives, Washington, D.C.*

FORT FETTERMAN
FROM C.G.COUTANT'S HISTORY OF WYOMING.

98. Fort Fetterman, Wyoming (1867). Established by Major William M. Dye. Aerial view drawing by M. D. Houghton. *Wyoming State Archives and History Department, Cheyenne.*

99. Fort Fred Steele, Wyoming (1868). Established by Major Richard Dodge. View from east-northeast; painting by Anton Schonborn. *Amon Carter Museum of Western Art, Fort Worth, Tex.*

protect railways, settlements, and emigrants en route from Minnesota to the gold fields of Montana. Since raids were feared, fortifications appeared.

Among those forts built for the safety of emigrants, as well as to protect the navigation of the Missouri River, was Fort Rice, North Dakota (1864), established by Brigadier General Alfred Sully (figs. 101, 102). The enclosure for this post consisted of two-inch-thick planks secured to a wooden framework. At diagonal corners were two-story blockhouses built of square logs dovetailed at the corners, with the upper story skewed forty-five degrees with respect to the lower. The first shelters were built of cottonwood logs, with roofs of "poles and slabs covered with earth."[30] Four years later they were replaced with balloon-framed buildings insulated with adobe between the studding.

More than one hundred miles northeast of Fort Rice, Fort Totten, North Dakota (1867), originally surrounded by a stockade, was built as part of a plan to confine the Indians of the area to a reservation and to protect emigrants (fig. 103). Named to honor Brigadier General Joseph G. Totten, chief engineer, who had died three years before its founding, it was also built of rough logs, which were subsequently replaced by permanent brick buildings, the clay for which was found near the site.

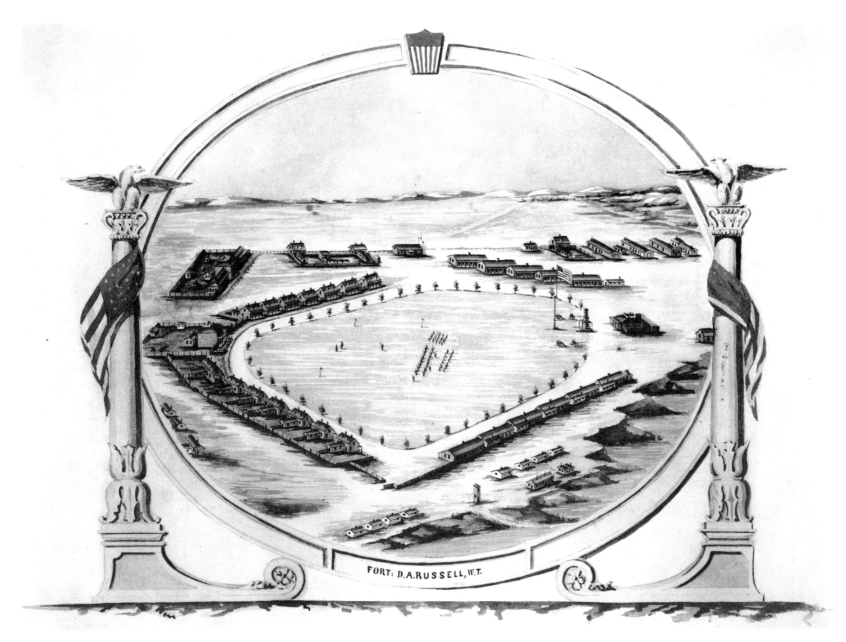

FORT. D.A.RUSSELL, W.T.

100. Fort D. A. Russell, Wyoming (1867). Established by Colonel John D. Stevenson. *Dr. and Mrs. Franz Stenzel.*

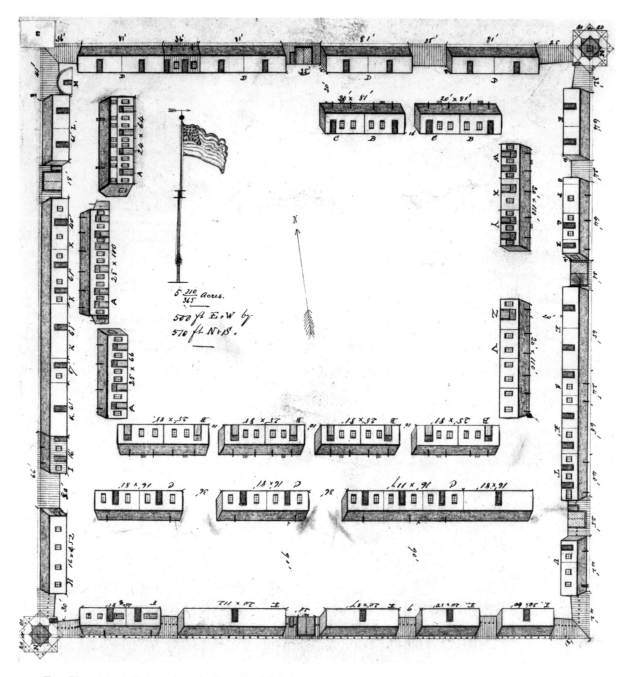

101. Fort Rice, North Dakota (1864). Plan. Established by Brigadier General Alfred Sully. *Minnesota Historical Society, Saint Paul.*

102. Fort Rice, North Dakota (1864). Painting (ca. 1870) by Seth Eastman. *Architect of the Capitol, Washington, D.C.*

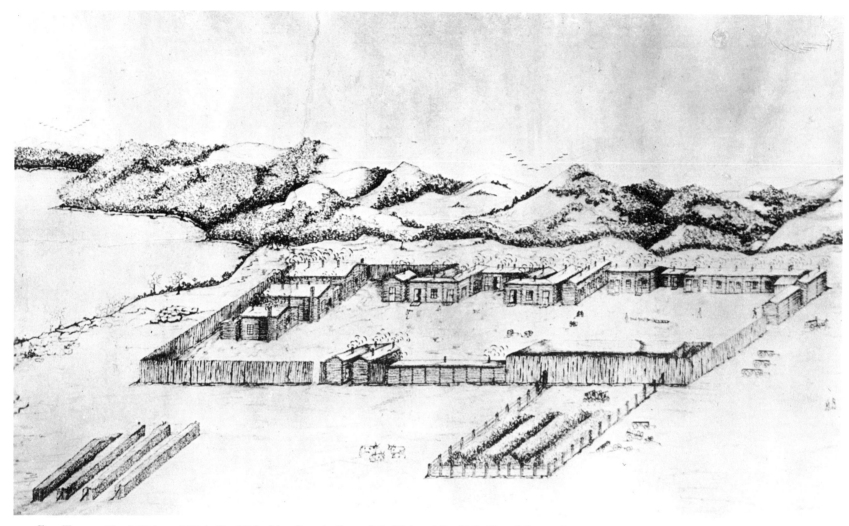

103. Fort Totten, North Dakota (1867). Established by Captain Samuel A. Wainwright. U.S. Signal Corps photograph. *National Archives, Washington, D.C.*

FORTS ON THE BOZEMAN TRAIL

The region in Dakota Territory which lay at the eastern foot of the Big Horn Mountains in the present states of Montana and Wyoming was among the most dangerous areas traversed by emigrants in the westward expansion. They faced depredation and death as they traveled the Bozeman Trail through territory recognized as belonging to the Indians by treaties with the Cheyenne, Arapaho, and Sioux tribes. However, this road along the east side of the Big Horn Mountains north of the North Platte River was one of the most direct routes to the gold fields of Montana.

In the mid-sixties, the army moved to protect Bozeman Trail travelers from hostilities. Overlooking the muddy waters of the Powder River, Fort Reno, Wyoming (1865), was erected as one of a chain of three forts constructed before the official abandonment of the trail in 1868 (fig. 104). The urgency of active defense in the Powder River country was suggested by the design of Reno. Buildings of cottonwood logs with earth roofs were mostly enclosed within a stockade, also of cottonwood, roughly rectangular in plan.[31] At each of the corners was located a flanking structure, three of which appear to have been hexagonal in form. Atop the three flankers rose offset, cubical watchtowers.

Like Reno, Fort Phil Kearny, Wyoming (1866), the most famous and interesting defense on the Bozeman Trail, was erected to protect northbound civilian trains (fig. 105). Colonel Henry B. Carrington selected the location on high ground near the Big Horns with the usual attention to natural resources. Adjacent to the site was the clear Little Piney Creek; good hay was found in nearby meadows; a short distance to the west there was an inexhaustible coal deposit; for building purposes, pine, fir, and spruce were also within a few miles to the west. Finally, Carrington saw in the site inherent advantages for defense. He reported that the plateau had "a natural glacis . . . falling off about 60 feet at an angle of 45°+. An engineer would hardly make a more perfect grade for the sweep of fire."[32]

Built while the soldiers lived in tents under constant threat of harrassment by Indians, the stockade of pointed logs with two sides hewn flat enclosed an area of approximately six hundred by eight hundred feet (fig. 106). On diagonal corners were located flanking blockhouses.

Enlisted men at Fort Phil Kearny constructed the buildings of pine and earth. Pine was a soft, workable wood easily hewn with axes and adzes, split with froes, or sawed at a horse- or steam-powered sawmill, both of which were used there. Typically, the walls were built with tenon and groove joints (fig. 107). This technique of construction—similar to a French type known as the "Red River frame" in Canada—was used on the barracks and was described by Frank M. Fessenden, a member of the band which accompanied Carrington's expedition into the wilderness. "We hewed squar post and cut a mortise the whole leinth and set them on the sill every twelve ft then cut logs the proper leingth hewed two sides cut tenent [tenoned] on each end and droped them in thease upright posts with the bark to the weather."[33] Carrington noted that tenon and groove construction provided "rapidity of erection, neatness, tightness."[34]

The roofs on the barracks were supported by small, closely spaced poles. According to Fessenden, "For the roof we cut poles about four inches through the proper leingth. cut the small nots of and laid them cloase together coverd them with corn sacks or grass then coverd with six inches of soil. That gave us good warm quarters. When it got forty below we were confortable."[35] Although earth roofs did provide good insulation, erosion from wind and rain required occasional repairs.

North of Fort Phil Kearny, Fort C. F. Smith, Montana (1866), established by Captain Nathaniel C. Kinney, enclosed

104. Fort Reno, Wyoming (1865). U.S. Signal Corps photograph. *National Archives, Washington, D.C.*

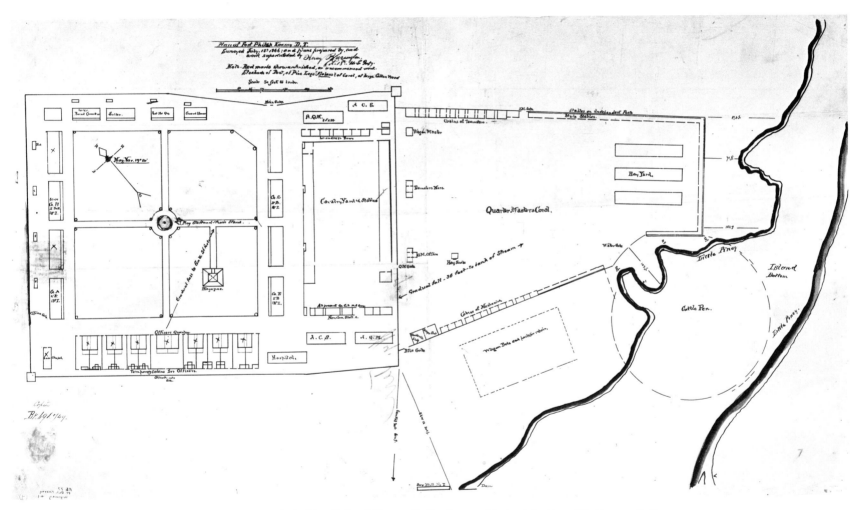

105. Fort Phil Kearny, Wyoming (1866). Plan. Established by Colonel Henry B. Carrington. *National Archives, Washington, D.C.*

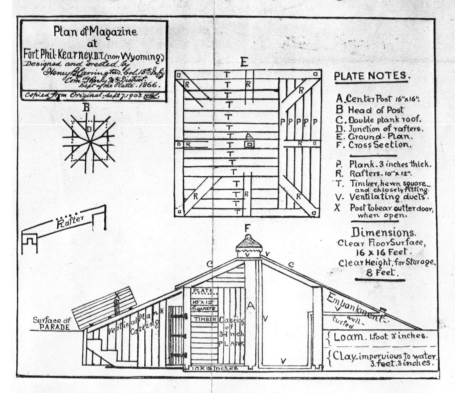
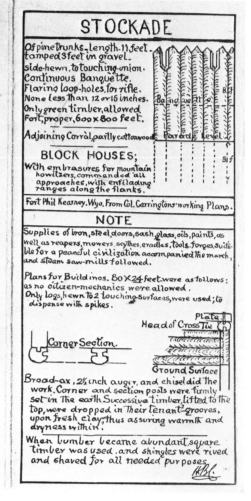

106. Fort Phil Kearny, Wyoming (1866). Details of magazine, stockade, and buildings (1903) drawn by Colonel Henry B. Carrington. *National Archives, Washington, D.C.*

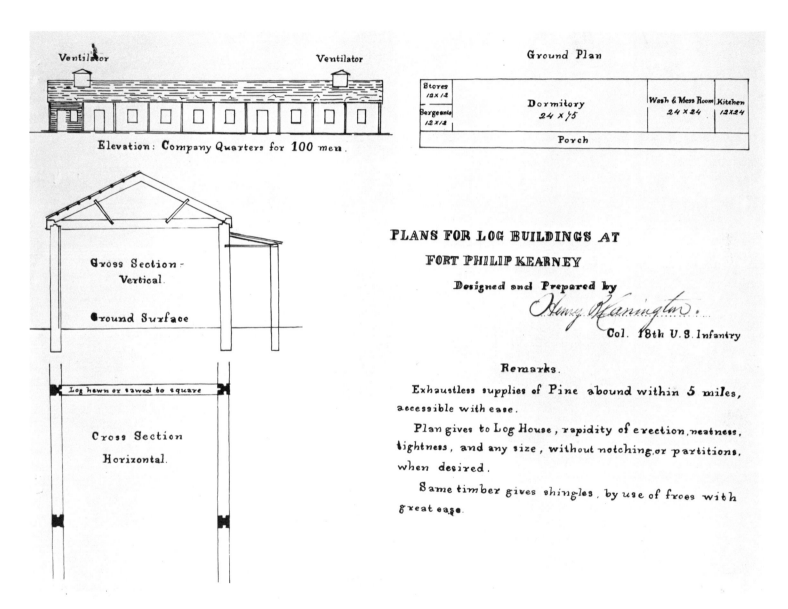

Ventilator

Ventilator

Elevation: Company Quarters for 100 men.

Cross Section -
Vertical.

Ground Surface

Log hewn or sawed to square

Cross Section
Horizontal.

Ground Plan

Stores 12x12				
Sergeants 12x12	Dormitory 24x75	Wash & Mess Room 24x24	Kitchen 12x24
	Porch		

PLANS FOR LOG BUILDINGS AT

FORT PHILIP KEARNEY

Designed and Prepared by

Henry Carrington

Col. 18th U.S. Infantry

Remarks.

Exhaustless supplies of Pine abound within 5 miles, accessible with ease.

Plan gives to Log House, rapidity of erection, neatness, tightness, and any size, without notching or partitions, when desired.

Same timber gives shingles, by use of froes with great ease.

107. Fort Phil Kearny, Wyoming (1866). Plans and details of log buildings (1866) drawn by Colonel Henry B. Carrington. *National Archives, Washington, D.C.*

108. Fort C. F. Smith, Montana (1866). Established by Captain Nathaniel C. Kinney. Perspective view; sketch by Captain I. D'Isay after a drawing by Anton Schonborn. U.S. Signal Corps photograph. *National Archives, Washington, D.C.*

a three-hundred-foot-square area (fig. 108). Built under the constant threat of clashes with Indians, the defensive wall was composed of adobe buildings and stockading. Unlike the other forts on the Bozeman Trail, it had bastions formed by upright logs located on diagonal corners. On one of the shoulder angles of each bastion was placed a bartizan from which the approaches to the fort could be watched.

In 1868, as a result of constant Indian pressure, the Bozeman road was closed. The three forts, which had been established at considerable expense in labor and life, were abandoned and destroyed by either troops or Indians. Architecturally, they had fulfilled their function, but the famed Sioux chief Red Cloud refused to negotiate for peace while they were in existence.

Indian Defense in the Northwest

While forts of logs and earth such as those along the Bozeman Trail reflected the primitive and unstable condition of the wilderness in which they were built, the appearance of buildings in distinct architectural styles symbolized the civilizing of other regions. On the frontier, energy was devoted to a more opulent architecture whenever pressures for active defense lessened. After realizing basic functional demands, builders turned their attention to decorating and glorifying architecture—a characteristic that has typified many epochs.

In the Northwest, the planning of military outposts for the control of Indians followed familiar patterns. Conditions motivating site selection and building organization on the site were the same as those that prevailed in other regions. The availability of resources influenced locations, while the nature of hostilities in a particular area influenced the character of fortifications. Then, as the region developed, various architectural styles appeared.

At Fort Dalles, Oregon (1850), new buildings that were built on a site formerly occupied by a civilian stockade called Fort Lee (1848) were designed in the Gothic Revival style—a mode which was prominent throughout the nineteenth century for ecclesiastical edifices and which become fashionable in the latter half of the century for secular buildings. Representative of the style was the officers' quarters with its steep-pitched roofs, window hoods, diamond window panes, and vertical board and batten siding as well as its irregular plan (fig. 109). This style was thought by some to be particularly appropriate for rural areas with picturesque landscapes.

The Gothic Revival also appeared at Fort Simcoe, Washington (1856), established by Major Robert Seldon Garnett, a West Point graduate. At this post, sixty-five miles north of The Dalles, the building housing the commanding officer (ca. 1857) was likewise built in the medieval-inspired mode. The picturesque style, contrasting with the more restrained design of other buildings on the post, emphasized the place of command, thereby fulfilling an additional function (fig. 110). Other board and battened buildings housing officers had attenuated Greek Revival details.

Other historic stylistic influences appeared in Montana. In the design of the buildings at Fort Assiniboine (1879), established by Colonel Thomas H. Ruger, there occurred another variation of the Gothic Revival. The two end buildings of officers' row, which flanked a series of Mansard-roofed buildings, were two stories high and built of brick. The parade-side lines of the low-pitched roofs were concealed behind a parapet. Forming a conspicuous part of these two end buildings were three-story towers with castellated parapets derived from medieval military architecture. Characteristic of nationwide trends at that time, the merlons which gave the towers their distinctive design were, of course, romantic rather than functional. Evidently the castellation that had been developed

Fort Dalles. Col. R.—

109. Fort Dalles, Oregon (1850). Established by Captain Stephen S. Tucker. Painting by William B. McMurtrie. *Boston Museum of Fine Arts*.

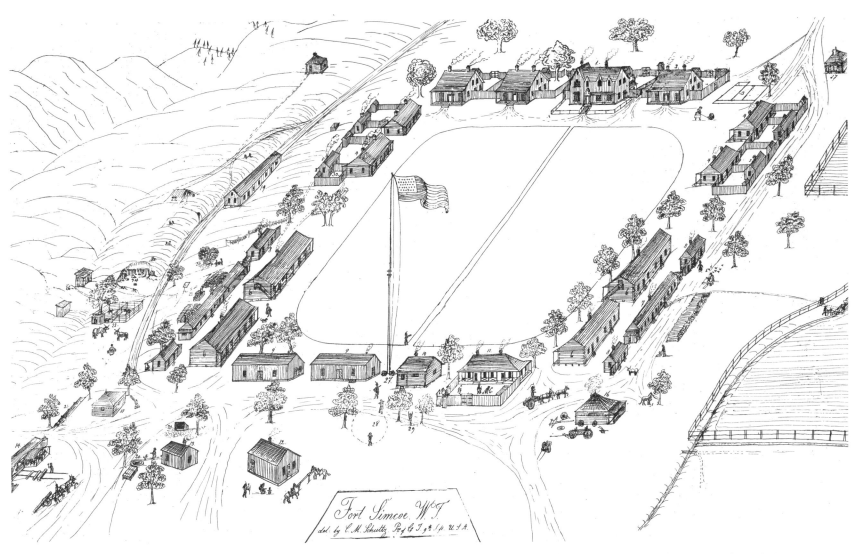

Fort Simcoe, W.T.
del. by C. M. Schultz Pvt Co I 9th Inf. U.S.A.

110. Fort Simcoe, Washington (1856). Established by Major Robert Seldon. Restoration of original drawing by Private C. M. Schultz, (1858). *Northwest Collection, University of Washington Libraries, Seattle.*

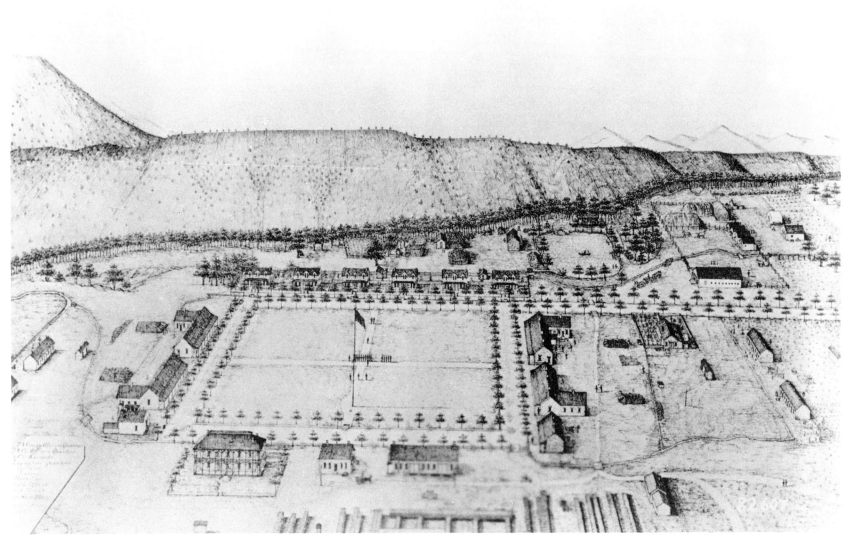

111. Fort Cameron, Utah (1872). Established by Major John D. Wilkins. *United States Military Academy, West Point.*

for the military architecture of antiquity was considered symbolically appropriate for expressing the military purpose of the buildings at Assiniboine.

Among the last nineteenth-century posts to be established in the West was Fort Cameron, Utah (1872). Characteristic of the last western forts, it was without fortifications and was laid out in rectangular form (fig. 111). Located around the parade were barracks and officers' quarters. Reflecting the region in which they were built, the buildings, except for the stables on the west, had walls of black basaltic lava from the nearby mountains. However, these durably constructed works were occupied by the military only until 1883, when the fort was sold at auction, evidently to representatives of the Church of Latter Day Saints.[36]

Thus, in the westward movement, while expressing varying degrees of stylistic and technical development, military forts had works for defense that were adjusted to varying conditions dependent upon their relationships with each particular tribe of Indians. The early fortifications planned for Indian country—for example, Fort Smith—were over-designed. Lacking practical knowledge of the West, military architects had evidently anticipated that Indians might lay siege to their forts, so they designed them accordingly. Experience ultimately proved that few architectural works specifically designed for active defense were needed in many areas. As a result, the open fort developed. Only in areas occupied by notoriously savage Indians did cautious military builders provide strong enclosures with blockhouses for flanking fire on assailants.

From the point of view of military function, then, most western forts were not designed to resist seige but rather to serve as bases from which Indians could be pursued and punished for depredations on settlers; others, such as Fort Union, served as bases for soldiers who escorted wagons through a wilderness occupied by unfriendly natives. However, while the posts provided bases from which to operate, their mere existence certainly must have been a deterrent to mischief.

In addition to their military role, the forts of the West provided numerous benefits for the development of civilization. The establishment of posts opened new roads and provided for the protection of energetic adventurers and expeditions as well as established settlers. Forts also served as bases where enterprising sutlers could bring commerce to the West, providing supplies and refreshments to soldiers as well as to emigrants. Posts like Fort Laramie provided supplies for wagon trains traveling the natural highways toward new frontiers. Some posts became stations for the pony express; still others, such as Fort Davis, were stagecoach stops for weary travelers. All of these functions, of course, suggest that the contributions of the forts to the civilizing and development of the West extended beyond patrol duty and Indian campaigns.

Through the establishment of military posts yet other contributions were made to the development of western culture. Many posts maintained libraries or reading rooms, and some—for example, Fort Davis—had schools. Post chapels provided a setting for religious services and weddings. At many points in the wilderness, post bands provided entertainment and boosted morale. During the last part of the nineteenth century, to reduce expenses, gardening was encouraged at the forts, thus making experimental agriculture another activity of the military. While Indian control was the prime concern, the military stationed at the various forts also played a role in civilian life. Soldiers assisted in maintaining order, and civilian officials often called upon the army for protection.

Certainly among other significant contributions the army made to the improvement of the conditions of life was the investigation of the relationships between health, climate, and architecture. From the earliest colonial times throughout the nineteenth century disease ranked as the foremost problem in defense. It slowed construction on forts and inhibited their military function. Official documents from many regions contained innumerable reports of sickness which virtually incapacitated entire garrisons. In response to the problems, detailed observations on architecture and climate and their relationships to the frequency of the occurrence of various diseases were recorded at various posts across the nation by military surgeons. The objectives of these records, along with statistics on sickness and mortality, were published in 1856 in a Congressional document the purpose of which, according to the surgeon general, was to "make the records . . . practically useful to the physician in *civil* life, and to render them subservient to the elucidation of the effects of climate in causation and development of disease."[37]

. . .

Throughout the history of America, architecture of defense has been characterized by beauty of form resulting from clarity of purpose. Fortifications, obviously significant in the political history of all periods, occupy important places as architectural examples of fulfillments of form and function concepts, and many rank among the most functional and beautiful works constructed before the twentieth century. Today the manner in which the principles regulated form in response to function in military architecture must continue to be inspiring.

Appendix

Remarks on Drawings

The subject of illustration of works for defense in North America is a study related to but somewhat apart from aspects of form and function. Therefore, a few separate remarks will emphasize some of the significant qualities of various types of drawings, representative examples of which appear in the text.

Many of the earliest drawings were primarily for purposes of general description and record, not for specific instructions to builders. Because of these intents and the limited skill of the artists, the drawings that resulted were abstract and pictorial rather than technical. Intuitive products of the artist, they did not conform to any conventionalized or systematic method of drawing. They, along with some written descriptions, are the only records of the first architecture observed or produced by white men during the colonial periods.

It is interesting to observe the techniques that were employed to communicate the essential features of three-dimensional forms. Early illustrators obviously sought to communicate as much information on a single drawing as possible. Although perspective would generally be the clearest method of creating the illusion of three dimensions, it was not used in some early drawings, either because of a lack of knowledge or because of a desired simplicity.

For example, sixteenth-century drawings of forts at Saint Augustine consisted of plan views of the defensive enclosure with side views of trees, buildings, and various other components of the fort superimposed. In this type of drawing, the elements shown in elevation were not to scale with the plan and were evidently symbolic.

Several early French illustrators likewise employed interesting techniques to indicate three dimensions. Champlain used a type of oblique drawing in the illustrations of his *habitations* and Indian fortifications. The scale of the various components of the drawings was again inconsistent.

As colonial competition over the North American continent became more intense, and fortifications became more formidable and scientific, drawings became more technical and detailed. Men trained as engineers often developed corresponding competence in drawing. The French engineer Chaussegros de Léry, for example, produced many fine plans and illustrated the manuscript of his treatise on fortification with many handsome sketches.

Throughout the nineteenth century there was a continued increase in the number of excellent engineer-delineators. Guillaume Tell Poussin, who had attended the École des Beaux-Arts in Paris and who was employed as a topographical engineer in the United States, worked on the permanent system of defense with Simon Bernard and drafted many fine drawings in color of existing and proposed works for the Gulf Coast. General J. G. Totten, chief engineer, also produced many finely delineated plans.

The importance of engineers who were skilled as draftsmen and delineators was recognized by the United States Military Academy when art training was developed in the curriculum. In the nineteenth century, courses in drafting, perspective, freehand sketching, and topographical rendering were required. Many graduates became excellent technical illustrators, and some became artists.

Drawings by military engineers made in connection with their official work were essentially of four types: plans for proposed works; plans showing progress and the stage of completion of fortifications under construction; as-built drawings of completed forts; and plans of captured works. Many versions of these different types were often made of the same fortifications. For all, the techniques of drawing were similar.

Because of the many advancing, receding, and inclined planes in plans of permanent fortifications, they were often difficult to read, particularly for the layman. Therefore, certain conventions and techniques were adapted by many engineers to improve the readability of drawings. In plan views, lines indicated changes of planes. Since fortifications often had forms that were characterized by many inclined surfaces, drawings were commonly quite complicated—the enceinte alone, for example, might have as many as nine different planes. The intersections of earth planes were drawn with fine black lines. Slopes less than 45 degrees were drawn with contour lines indicating one-yard vertical intervals, commencing from the plane of the site. For clarity, crests of parapets were indicated by heavy black lines; the line of the cordon, which was the base or master line of the works, was drawn with a wide red line.

Washes of color were often used to further distinguish between various parts of the works. Representing different materials with their characteristic hues, their use was logical and functional. In addition to indicating the trace of the cordon, red was used to illustrate masonry. Blue, or blue-green, represented wet ditches and natural bodies of water. To delineate grass slopes, shades of green were applied. Tints of burnt sienna, umber, or ochre were used to indicate dry ditches and other sand or earth works. However, such

level parts as banquettes and terrepleins were generally left white. To represent iron, and roofs of buildings, indigo was employed. Yellow was often applied to show wood, but it was also used to show unfinished structures. These same colors were used to delineate drawings of profiles and sections.

To create the impression of sloping planes, colors were varied in value. While terrepleins and banquettes were usually white, height in other works was represented by dark values and depth by light values, thereby creating an illusion of depth.

The delineation of light, shades, and shadows increased the definition of depth and form. The direction of light was customarily assumed to come from the upper left-hand corner of the sheet on which the drawing was made. Washes, deep in value, were then applied to represent shadows. Planes facing the light were high in value, while those away from the direction of light were darker.

Conventions in shades and shadows, and in assumed directions of light, were also followed in the drawing of profiles. For masonry there was often a thick red line on the edge away from the assumed light source.

Although plans of fortifications were technical, beautiful lettering and decoration often appeared on the drawing, demonstrating the pride that engineers had in their work.

Graphical work on fortifications was not limited to the work of engineers. In the nineteenth century, architecture for defense was a popular subject for amateur and professional artists. Brigadier General Seth Eastman, after retiring from the army, created beautiful paintings of seventeen principal military works between 1870 and 1875. A part of the collection in the United States Capitol, these paintings were of diversely located subjects. Conrad Wise Chapman executed some fine paintings of fortifications in Charleston Harbor, South Carolina, which are now in the collection of the Confederate Museum. The setting of many plains forts was captured in a series of watercolors by Anton Schonborn which are in the Amon Carter Museum of Western Art. Soldiers stationed at various posts often produced sketches of their surroundings.

Drawings of fortifications are deposited in many different institutions in Europe and North America. Numerous records of colonial forts are in the Archivo General de Indias, Seville; the Servicio Histórico Militar, Madrid; the Bibliothèque Nationale, Paris; the Archives Nationales, Paris; the Public Records Office, London; and the British Museum, London. The Public Archives of Canada, Ottawa, likewise contains many drawings. The John Carter Brown Library, Providence, and the William L. Clements Library, Ann Arbor, have fine collections of material on early Americana. The National Archives, Washington, has approximately 25,000 sheets of plans and maps on American fortifications. Thus, to the historian of military architecture there is available a wealth of manuscript drawings.

Notes

INTRODUCTION

1. Henri Pirenne, *Medieval Cities: Their Origins and the Revival of Trade*, trans. Frank D. Halsey (New York, 1925), p. 40.

2. Flavius Josephus, *The Works of Flavius Josephus*, trans. William Whiston (New York, 1815), vol. 7, p. 5.

3. Marcus Vitruvius Pollio, *The Ten Books on Architecture*, trans. Morris Hicky Morgan (New York, 1960), p. 305–8.

4. Eugène Emmanuel Viollet-le-Duc, *Military Architecture*, p. 9.

5. Vitruvius, *Ten Books on Architecture*, p. 22.

6. Edward Gibbon, *The Decline and Fall of the Roman Empire* (London, 1910), vol. 1, p. 16.

7. Gibbon, *Decline and Fall*, p. 16.

8. John Womack Wright, *The Development of the Bastioned System of Fortifications, 1500–1800*, pp. 6, 34.

9. Viollet-le-Duc, *Military Architecture*, p. 204.

10. Various histories give different intepretations of the invention of the bastion. Charles James, *A New and Enlarged Military Dictionary . . . of the Different Systems of Fortification . . .*, observed, "Those who wrote on fortification 200 years ago, seem to suppose that bastions were a gradual improvement in the ancient method of building, rather than a new thought, that any one person could claim the honor of. It is certain, however, that they were well known soon after the year 1500. . . ." A. F. Lendy, *Treatise on Fortification; or, Lectures Delivered to Officers Reading for the Staff*, p. 490, wrote, "There is no certainty as to the precise date of the invention of bastions. . . ." Another nineteenth-century military source, Henry Wager Halleck, *Elements of Military Art and Science . . .*, 3d ed. (New York and London, 1862), p. 328, stated, "It is not known who changed the ancient towers into bastions. Some attribute it to an Italian . . . for a bastion was built at Turin as early as 1461. . . . It is most probable that the transition from the tower to the bastion was a very gradual one, and that the change was perfected in several countries at about the same time." Viollet-le-Duc, *Military Architecture*, p. 237, attributes to Sanmicheli a new bastion form at the city of Verona. A recent history, Bertrand Gille, *Engineers of the Renaissance*, p. 212, also states that the first perfect example seems to have been ". . . the bastion of the Maddelena, at Verona, built by San Michele [Sanmicheli] in the year 1527." However, Sigfried Giedion, *Space, Time and Architecture*, 5th ed. rev. (Cambridge, 1967), p. 43, states that bastions ". . . are said to have been invented by Francesco di Giorgio."

11. Triangular schemes were rarely used because they were difficult to defend; gorges were very narrow, and bastion faces were excessively long. One of Vauban's maxims stated that it was impossible to fortify a triangle.

12. Eugène Emmanuel Viollet-le-Duc, *Annals of a Fortress*, trans. Benjamin Bucknall (Boston, 1876), p. 279.

13. The tenaille arrangement formed a basis for systems proposed by several engineers after the sixteenth century. In the eighteenth century the French engineer Marc René, Marquis de Montalembert, developed a tenaille system, as did Lazare Nicolas Marguerite Carnot, another Frenchman. As late as 1866 an advocate for the system appeared. See F. P. J. Piron, "The Systems of Fortification Discussed and Compared," *United States Service Magazine*, 5 (Jan., 1866), 34–40; (Feb., 1866), 108–14; (March, 1866), 225–34; (April, 1866), 328–32.

14. See Daniel Halévy, *Vauban: Builder of Fortresses*. Vauban was born in Morvan, France. In 1651 he entered the service as a cadet. Only four years later, at the age of twenty-two, he received a commission as engineer-in-ordinary to the king. In 1678 Vauban was appointed inspector-general of the

fortification of France, a position which kept him constantly traveling. Four years before his death in Paris he was accorded the high honor of being appointed marshal of France.

15. *Biography universelle ancienne et moderne* . . ., new ed., s.v. "Vauban, Sébastien Le Prestre de."

16. Sébastien Le Prestre de Vauban, *De l'Attaque et de la defense des places*, preface.

17. The principles are from, respectively, maxims I, X, XV, and II in Sébastien Le Prestre de Vauban, *The New Method of Fortification as Practiced by Monsieur Vauban . . . Made English*, pp. 22–23.

18. See Wright, *The Development of the Bastioned System*, p. 157, and D. H. Mahan, *An Elementary Course of Military Engineering*, vol. 2, p. 46. The second system was used only twice—at Belfort (1684) and at Landau (1688)—whereas the third system was used only at New Brisac (1698).

19. Viollet-le-Duc, *Annals of a Fortress*, p. 308.

20. Vauban, *New Method of Fortification*, p. 22, maxim III.

1. COLONIAL FORTIFICATIONS

1. René Laudonnière, "The Second Voyage into Florida," in Richard Hakluyt, *The Principal Navigations, Voyages, Traffiques and Discoveries of the English Nation*, vol. 3, p. 392.

2. See Verne E. Chatelain, *The Defenses of Spanish Florida, 1565 to 1763*, pp. 41–55. The elements, war, and rebellion took a heavy toll. About a year after its construction, hostile Indians burned the first fort (1565); the second (1566) lasted less than a year, the sea encroaching upon it; the third (1566) was burned by the Spanish soldiers themselves after five years. Of the works built during the sixteenth century, the seventh (1586) endured the longest, remaining until it was destroyed by a storm in 1599.

The terms *stockade* and *palisade* are now often used interchangeably by historians. However, writers on fortifications distinguish between the two modes of construction. Palisades consist of pales spaced six to eight inches apart, while stockades are timber walls constructed from logs placed tightly together.

3. Chatelain, *Defenses of Spanish Florida*, p. 43.

4. Thomas Cates, "A summarie and true discourse of Sir Francis Drakes West Indian voyage, begun in the year 1585," in Hakluyt, *Principal Navigations*, vol. 4, pp. 24–25.

5. See Jeanette Thurber Connor, "The Nine Old Wooden Forts of St. Augustine," *Florida Historical Quarterly*, 4 (1926), 103–11, 171–80, for a history of these forts.

6. Drawing titled, "Plano del Fuerte Biejo que Está en San Augustin" (ca. 1593), fig. 2, in the Archivo General de Indias, Seville. Translation cited in Chatelain, *Defenses of Spanish Florida*, p. 54.

7. Governor Andres Rodriquez de Villegas to Philip IV, January 18, 1631, in Chatelain, *Defenses of Spanish Florida*, p. 146.

8. Albert C. Manucy, *The Building of the Castillo de San Marcos*, p. 8. However, the decision to build a stone fort had been made as early as 1595; see Chatelain, *Defenses of Spanish Florida*, p. 54.

9. Chatelain, *Defenses of Spanish Florida*, p. 64; Manucy, *Building of the Castillo de San Marcos*, p. 9.

10. J. C. Harrington, Albert C. Manucy, and John M. Goggin, "Archeological Excavations in the Courtyard of Castillo de San Marcos, St. Augustine, Florida," *Florida Historical Quarterly*, 34 (Oct., 1955), 102.

11. Hugh Morrison, *Early American Architecture from the First Colonial Settlements to the National Period* (New York, 1952), p. 182.

12. Chatelain, *Defenses of Spanish Florida*, p. 85.

13. Stanley Faye, "Spanish Fortifications of Pensacola, 1698–1763," *Florida Historical Quarterly*, 20 (1941), 153.

14. Information from print D/1251, Map Division, Public Archives of Canada, Ottawa.

15. "Narrative of the Expedition Made by Order of Louis XIV., King of France, under Command of M. d'Iberville, to Explore the Colbert (Mississippi) River and Establish a Colony in Louisiana," in *Historical Collections of Louisiana and Florida*, ed. and trans. B. F. French, 2d ser. (New York, 1875), p. 112.

16. "*Ce fort est de bois, à quatre bastions; deux sont de pièce sur pièce d'un pied et demi, d'un pied de haut, ponté comme un navire, sur quoy est le canon avec une parapet de quatre piede de haut; les deux autres de bonne palissade bien doublée. . . .*" D'Iberville au Ministre de la Marine, June 29, 1699, in Pierre Margry, *Découvertes et établissements des français dans l'ouest et dans le sud de l'Amérique septentrionale (1614–1754)* (Paris, 1879–88), vol. 4, p. 125.

17. ". . . *quatre bastions, de pièces sur pièces. . . .*" "Journal du Sieur D'Iberville," in Margry, *Découvertes et établissements*, vol. 4, p. 512.

18. See "Census of Louisiana by Nicholas de la Salle," *Mississippi Provincial Archives, French Dominion*, ed. and trans. Dunbar Rowland and Albert Godfrey Sanders (Jackson, Miss., 1929), vol. 2, p. 18.

19. See *Mississippi Provincial Archives*, vol. 2, p. 392.

20. ". . . *construit en brique & fortifié à quatre bastions, . . . avec des demilunes, un bon fossé, un chemin convert & un glacis. . . .*" M. Dumont, *Memoires historiques sur la Louisiane* (Paris, 1753), vol. 2, p. 79. For drawings of the fort made in 1725 see Willard B. Robinson, "Military Architecture at Mobile Bay," *Journal of the Society of Architectural Historians*, 30 (1971), 122.

21. For a drawing by the French topographical engineer G. T. Poussin, see Record Group 77, Drawer 81, Sheet 4, Cartographic Branch, National Archives, Washington, D.C.

22. H. Mortimer Favrot, "Colonial Forts of Louisiana," *Louisiana Historical Quarterly*, 25 (1946), 736–37.

23. Philip Pittman, *The Present State of the European Settlements on the Mississippi*, p. 89.

24. Peter J. Hamilton, *Colonial Mobile: An Historical Study*, 2d ed. (Boston, 1952), p. 96.

25. For a plan of Fort Tombecbé, see Marc de Villiers du Terrage, *Les dernières années de la Louisiane française* (Paris, 1903), p. 171.

26. See J. A. de la Peña, *Derrotero de la expedición en la Provincia de los Texas...*, drawings following pp. 20 and 27 for plans and descriptions.

27. Robert S. Weddle, *The San Sabá Mission: Spanish Pivot in Texas* (Austin, 1964), p. 148.

28. Rexford Newcomb, *The Old Mission Churches and Historic Houses of California* (Philadelphia, 1925), p. 66.

29. *Ibid.*

30. Cited in Harrie Rebecca Piper Forbes, *California Missions and Landmarks: El Camino Real*, 8th ed. (Los Angeles, 1925), p. 250.

31. See Newcomb, *Old Mission Churches and Historic Houses*, p. 66.

32. Drawing of Fort Duquesne, Record Group 77, Drawer 145, Sheet 19, Cartographic Branch, National Archives, Washington, D.C. Drawing based on an English prisoner's report.

33. Charles Morse Stotz, "Defense in the Wilderness," in *Drums in the Forest* (Pittsburgh, 1958), p. 132. Caesar described a similar type used by the Celts. See Viollet-le-Duc, *Military Architecture*, p. 4.

34. Edward P. Hamilton, *Fort Ticonderoga: Key to a Continent*, p. 39.

35. De Lotbinière to Marquis de Vaudreuil, October 31, 1756, in S. H. P. Pell, *Fort Ticonderoga; A Short History Compiled from Contemporary Sources*, pp. 23, 25.

36. See William Hunter, *Forts on the Pennsylvania Frontier, 1753–1758*, p. 113.

37. Hamilton, *Fort Ticonderoga*, p. 50.

38. *Ibid.*, p. 89.

39. Hunter, *Forts on the Pennsylvania Frontier*, p. 226.

40. For a sketch, see *ibid.*, p. 237.

41. Captain Eyre to Major General Shirley, September 10, 1755, M.G. II, C.O. 5, Vol. 16–1, Public Archives of Canada, Ottawa.

42. Washington to John Robinson, August 5, 1756, in *The Writings of George Washington: Being His Correspondence, Addresses, Messages, and Other Papers, Official and Private*, ed. Jared Sparks, vol. 2, p. 171.

43. Francis Parkman, *France and England in North America* (New York, 1965), vol. 8, *Montcalm and Wolf*, p. 495.

44. For example, Fort Duquesne was approximately 160 feet on a side, Castillo de San Marcos varied between about 300 and 320 feet, and Fort William Henry was 115 feet on its longest side.

45. Originally, a rectangular, bastioned fort measuring 450 feet by 550 feet between bastion points was proposed. This plan was superseded by the pentagonal fort because it was believed that the latter fitted the ground better. See drawings H4/950.25 and H4/950, Public Archives of Canada, Ottawa.

46. The sites of Fort Rascal (1755–56)—also called Fort George—and Fort Oswego (1727) were on opposite sides of the river. Old Fort Ontario (1755) was situated approximately where the new fort was constructed. Forts Rascal and Ontario were weak, stockaded enclosures, while Oswego was an indefensible stone and earth work.

47. James Oglethorpe to the Trustees for the Founding of the Colony, December 29, 1739, in *Collections of the Georgia Historical Society*, vol. 3, p. 100.

48. Albert C. Manucy, *The Fort at Frederica*, Notes in Anthropology, vol. 5, pp. 9, 33.

49. Harman Verelst to Thomas Causton, August 11, 1737, in Amos Aschback Ettinger, *James Edward Oglethorpe: Imperial Idealist*, p. 241.

50. See Samuel Wilson, Jr., "Colonial Fortifications and Military Architecture in the Mississippi Valley," in *The French in the Mississippi Valley*, ed. John Francis McDermott, pp. 111–12.

51. Villiers du Terrage, *Les Dernières années de la Louisiane française*, p. 20. The work was directed by Perier.

52. Dumont, *Memoires historiques sur la Louisiane*, vol. 2, p. 52.

53. Villiers du Terrage, *Les Dernières années de la Louisiane française*, p. 107.

54. Jack D. L. Holmes, "Some French Engineers in Spanish Louisiana," in *French in the Mississippi Valley*, p. 127.

55. Wilson, "Colonial Fortifications and Military Architecture," p. 113.

2. Transitional Work

1. Hamilton, *Fort Ticonderoga*, p. 102.

2. *Ibid.*

3. Stotz, "Defense in the Wilderness," p. 174.

4. Claud H. Hultzén, Sr., *The Story of an Ancient Gateway to the West: Old Fort Niagara*, p. 48.

5. *Dictionary of American Biography*, s.v. "Gridley, Richard." Richard Gridley studied military engineering under John Henry Bastide, a British officer who contributed to the development of fortifications for Boston Harbor. In 1746 Gridley planned fortifications for that harbor in preparation for an attack expected from the French.

6. Drawing 135, Section Outre-mer, Archives Nationales, Paris.

7. William Wood and Ralph Henry Gabriel, *The Pageant of America*, vol. 6, *The Winning of Freedom*, p. 174.

8. Statement of General John Burgoyne, in John Fiske, *The American Revolution*, 2 vols. (Boston, 1919), vol. 1, p. 280.

9. Washington to the President of Congress, July 10, 1775, in *The Writings of George Washington*, vol. 3, p. 18.

10. Washington to the Committee of Safety of Pennsylvania, June 17, 1776, in *The Writings of George Washington*, vol. 3, p. 427.

11. *Biographie universelle ancienne et moderne*, s.v. "Duportail." Duportail, a graduate of the technical school at Mézières, France, was the chief of the American engineers and rose to the rank of brigadier general. After the Revolution he returned to France, where he became minister of war. See Elizabeth S. Kite, *Brigadier-General Louis Lebèque Duportail* (Baltimore, 1933), pp. 7–9. In 1792 he fell into political disfavor and was able to escape execution by fleeing to America. A decade later he boarded a ship to return to France but died on the voyage.

12. Peter J. Guthorn, *American Maps and Map Makers of the Revolution* (Monmouth Beach, N.J., 1966), p. 23. Gouvoin was born in Toul, France, and studied art at the technical school in Mézières. Among his contributions to the American cause were the fortifications at Verplank's Point, New York.

13. Kite, *Brigadier-General Duportail*, p. 46. General Howe was criticized for his failure to attack the camp. He defended his action by stating that the fortifications were too strong.

14. See *The Writings of George Washington*, vol. 5, footnote, p. 142; and Kite, *Brigadier-General Duportail*, p. 91. Kościuszko was appointed an engineer in October, 1776. Before his West Point engagement he had served at Ticonderoga and Mount Independence. He also planned the encampment for the American army at Bemis Heights. Guthorn, *American Maps and Map Makers*, p. 25. From the academy in Warsaw, Kościuszko received his education in the military sciences. In addition, he studied for a year at the Academie Royal de Peinture et de Sculpture in Paris.

15. General Putnam to the Commander-in-Chief, February 13, 1778, in *The Writings of George Washington*, vol. 5, p. 225.

16. John H. Mead, "Archeological Survey of Fort Putnam and other Revolutionary Fortifications at West Point, N.Y., 1967–68," unpublished study, United States Military Academy Museum, West Point, N.Y., pp. 76–79.

17. Jonathan Williams, "General Return of Fortifications from Massachusetts to Pennsylvania (1802)," Record Group 77, Drawer 245, Sheet 6, National Archives, Washington, D.C.

18. James Ripley Jacobs, *The Beginnings of the U.S. Army, 1783–1812* (Princeton, 1947), p. 282.

19. Louis Lebèque Duportail, "Memorial on the Works Made in the Highlands," September 13, 1778, in Kite, *Brigadier-General Duportail*, p. 100.

20. From a map in Joseph F. W. des Barres, *The Atlantic Neptune Published for the Use of the Royal Navy of Great Britain*, vol. 2.

21. John Johnson, "The Old Fort at Dorchester, S.C.," *South Carolina Historical and Genealogical Magazine*, 6, (1905), 128.

22. Captain Henry Bird to Brigadier General Powell, August 13, 1782, in Philip P. Mason, *Detroit, Fort Lernoult, and the American Revolution*.

23. Mason, *Detroit, Fort Lernoult, and the American Revolution*.

24. *Dictionary of American Biography*, s.v. "Montrésor, John." Montrésor, the son of another British military engineer, James Gabriel Montrésor (1702–76), was born at Gibraltar. He was brought to America by his father in 1754 and evidently acquired his engineering skill from his father and in the field. He was present at Amherst's siege of Louisbourg and Wolfe's siege of Quebec. Later, he worked on fortifications or barracks at Detroit, Boston, and New York.

25. Forbes, *California Missions and Landmarks*, p. 252.

26. "Instructions to Temporary Engineers," *American State Papers: Military Affairs*, vol. 1, pp. 71–108. The following were the appointed temporary engineers: Béchet Rochefontaine, John Jacob Ulrich Rivardi, Pierre Charles L'Enfant, Charles Vincent, Daniel Niven, John Vermonnet, Nicholas Francis Martinon, and Paul Hyacinte Perrault. The legislation, approved March 20, 1794, was "An Act to Provide for the Defence of Certain Ports and Harbors in the United States," *Debates and Proceedings in the Congress of the United States . . .*, 3d Cong., 1834, vol. 4, pp. 1423–24.

27. "Instructions to Temporary Engineers," pp. 73, 83, 87, 93, 95, 101.

28. Rochefontaine, "A General Return of the Situation of the Fortifications of the Seaport Towns in the States of New England," *American State Papers: Military Affairs*, vol. 1, p. 76.

29. "Report on Fortifications," *American State Papers: Military Affairs*, vol. 1, p. 120.

30. Williams, "General Return of Fortifications from Massachusetts to Pennsylvania."

31. Joseph G. Totten, *Report on Fortifications*, 32d Cong., 1st sess., 1851, House Exec. Doc. No. 5, pp. 90–91.

32. Timothy Pickering, "Report of the Department of War, Relative to the Fortification of the Ports and Harbors of the United States," *Annals of Congress*, vol. 6, p. 2572.

33. Williams, "General Return of Fortifications from Massachusetts to Pennsylvania."

34. See *American State Papers: Military Affairs*, vol. 16, pp. 72, 87–88. Rivardi was ordered to fortify Norfolk, Virginia. In 1794 plans were submitted to the secretary of war for Fort Nelson and Fort Norfolk, which were on the opposite side of the river.

35. "Report of Lewis [sic] Tousard to James McHenry," October 12, 1798, Record Group 77, Entry 20, File G, No. 135, National Archives, Washington, D.C.

36. *American State Papers: Military Affairs*, vol. 1, p. 309. For a plan of this

early work, see Record Group 77, Drawer 27, Sheet 1, Cartographic Branch, National Archives, Washington, D.C.

37. Record Group 77, Drawer 36, Sheet 16, Cartographic Branch, National Archives, Washington, D.C. The drawing is dated 1801. Mangin also designed the New York prison.

38. Richard Walsh, "The Star Fort: 1814," *Maryland Historical Magazine*, 54 (September, 1959), 296.

39. For the configuration of Fort Whetstone, see Antoine Pierre Folie, engraved map (1792), John Carter Brown Library, Providence, R.I.

40. J. J. U. Rivardi to the Secretary of War, *American State Papers: Military Affairs*, vol. 1, p. 88.

41. Louis de Tousard was born in Paris. After graduating from artillery school in Strasbourg, he served in the American army from 1777 until 1778, when he was wounded. In 1793 and in 1802 he returned to France for two and three years, respectively. Following each stay he journeyed back to the United States. Tousard was the author of *American Artillerist's Companion; or, Elements of Artillery*.

42. Lee H. Nelson, *An Architectural Study of Fort McHenry*, pp. 13, 21.

43. The development of casemates has a long history. Although the etymology is not certain, according to a prominent American military engineer, John G. Barnard, the term is from the Spanish *casa-mata*, a compound meaning "house" and "to kill," but it may also have designated a low or hidden house. The term was perhaps applied first by Italian and German engineers in connection with tiers of cannons flanking ditches. In any event, by the nineteenth century a casemate was a vaulted room employed to protect munitions, garrisons, and artillery.

The use of casemates can be traced back to the early part of the sixteenth century when both the German artist Albrecht Dürer and the Italian engineer Michele Sanmicheli employed them. Later, Louis XIV's engineer, Vauban, included tower bastions with casemates as parts of his second and third systems of fortification. However, late in the eighteenth century it was another French engineer, the Marquis de Montalembert, who placed the most importance on casemates. Cf. John G. Barnard, "Eulogy on the Late Joseph G. Totten," *Annual Report of the Smithsonian Institution*, 39th Cong., 1st sess., 1866, House Exec. Doc. No. 102, pp. 157–58. R. Delafield, *Report of Colonel R. Delafield, U.S. Army, on the Art of War in Europe in 1854, 1855 & 1856*, 36th Cong., 1st sess., 1860, Senate Exec Doc. No. 59, p. 177.

44. For theory on this military architect's work, see Marquis de Montalembert, *La Fortification perpendiculaire ou essai sur plusieurs manières de fortifier la ligne droite, le triangle, le quarré et tous les polygônes, de quelquéntendue qu'en soient les côtés, en donnant á leur défense une direction perpendiculaire.*

45. *Ibid.*, vol. 2, p. 316; vol. 3, p. 20.

46. Drawing: New York, No. 2 (ca. 1675), Colonial Office Library, London; copy in John Carter Brown Library, Providence, R.I.

47. In 1822 its use as a harbor defense was discontinued. It was used thereafter for various other functions. It served as an ornamental garden, theater, immigrant depot, and aquarium. The shoreline was moved out beyond it so that it is now located on land at the Battery.

48. Report of Secretary W. Eustis, *American State Papers: Military Affairs*, vol. 1, p. 31.

49. *Dictionary of American Biography*, s.v. "L'Enfant, Pierre Charles." Pierre Charles L'Enfant was born in Paris, the son of a painter, Pierre L'Enfant. He came to the United States in 1777 with Colonel du Coudray and served at Valley Forge. In 1778 he was commissioned captain of engineers. During the Revolution he also served at Savannah and Charleston.

50. Major L'Enfant to the Secretary of War, *American State Papers: Military Affairs*, vol. 1, p. 83.

3. The Permanent System

1. "Message to Congress on the Re-examination of Positions on Dauphin Island and Mobile Point for Fortifications," *American State Papers: Military Affairs*, vol. 2, p. 368.

2. Guillaume Tell Poussin, *Les Etats-Unis d'Amerique*, p. 77.

3. See *Nouvelle biographie générale*, vol. 5, s.v. "Bernard, Simon"; and *Biographie universelle ancienne et moderne*, s.v. "Bernard, Simon." Simon Bernard was born in Dôle, France. Although very poor, he entered the École Polytechnique in Paris in 1794. In 1797 he joined the Corps du Genie. Fortifications under his direction were begun in 1806 in Trieste and Ragusa, and in 1809 he worked on the fortifications of Antwerp, Belgium. Bernard was made aide-de-camp of Napoleon I in 1813 and was with him at the Battle of Waterloo. While working in the United States, he held the rank of brigadier general.

4. See John G. Barnard, "Eulogy on the Late Joseph G. Totten," pp. 150–51; *Dictionary of American Biography*, s.v. "Totten, Joseph G." George W. Cullum, *Biographical Register of the Officers and Graduates of the U.S. Military Academy from 1802 to 1867*, 2 vols. (New York, 1879), vol. 1, p. 94. J. G. Totten was a native of New Haven, Connecticut. He attended the United States Military Academy from 1802 to 1805, becoming the tenth graduate. His active career as a military engineer commenced under the supervision of Jonathan Williams in New York Harbor, where he was employed on the construction of Castles Williams and Clinton. An inquisitive scientist, Totten conducted many experiments on building materials, some of which were published. He visited most, if not all, of the permanent forts in the country which were under construction in the first half of the nineteenth century, making observations on their construction.

5. The board of engineers, constituted November 16, 1816, originally consisted of Brigadier General Bernard, Colonel William McRee, and

Lieutenant Colonel Totten. Bernard was president of the board. In 1817 Totten was replaced by Brigadier General Joseph Gardiner Swift (1783–1865). Bernard's work created animosity among the American engineers. Due to disagreements, Swift, who was chief engineer from 1812 until 1818, resigned from the board and from the army in 1818. McRee also resigned, apparently because of views similar to those of Swift. In 1819 Totten was reappointed, and the permanent board thereafter consisted of Bernard and Totten. However, Captain J.D. Elliot of the navy served in 1821–22, and officers superintending construction on individual fortifications were named as *ex officio* members of the board.

6. There were numerous manuscript, as well as published, reports relating to the objectives of the permanent system. See, for example, "Report of Defences of the U.S. Frontiers, 1822," and "Report on Defenses of the Seacoast, March 24, 1846," Record Group 77, Corps of Engineers, Entry 223, National Archives, Washington, D.C.

7. "Report of the Board of Engineers," *American State Papers: Military Affairs*, vol. 2, p. 310.

8. "Revised Report of the Board of Engineers on the Defense of the Seaboard," *American State Papers: Military Affairs*, vol. 3, p. 284.

9. "Report of the Board of Engineers," *American State Papers: Military Affairs*, vol. 2, p. 308.

10. See Poussin, *Les Etats-Unis d'Amerique*. Born in 1794, Guillaume Tell Poussin was a descendant of a family of distinguished painters. His father was "laureat de l'academie des beaux-arts de France," and the distinguished French painter Nicolas Poussin was his uncle. Poussin arrived in the United States in 1814 and in 1816 worked under Benjamin Henry Latrobe as an inspector of art, sculpture, and decoration on the national capitol. Early in 1817 he was appointed assistant engineer in the Engineer Corps with a rank of captain. He was associated with Bernard in the United States approximately fifteen years. He produced a large number of beautifully drawn maps and plans for fortifications, most of which are in the National Archives, Washington, D.C. In 1831 he returned to Europe.

11. F. P. Blair, Jr., *Permanent Fortifications and Sea-Coast Defences*, 37th Cong., 2d sess., 1862, Rept. No. 86, p. 4.

12. Cf. Guillaume Henri Dufour, *De la Fortification permanente*.

13. "Report on the Reexamination of Dauphin Island and Mobile Point," *American State Papers: Military Affairs*, vol. 2, p. 373.

14. The work is often erroneously called Fortress Monroe instead of Fort Monroe. The term *fort* applies to structures for defense which contain only military garrisons. *Fortress* is a term designating a fortified town. Although Monroe was originally called a fortress, the prefix was changed officially in 1832.

15. Fort Monroe had a perimeter circuit of 2,304 yards compared to 308 for Forts Wood and Pike and 675 for Fort Morgan. It covered sixty-three acres.

16. Barnard, "Eulogy on the Late Joseph G. Totten," p. 150.

17. Tenailles with flanks had been used by Vauban, but they were without casemates. The concept of the double land front was later proposed for other works, including Fort Schuyler, New York (1883), at Throg's Neck.

18. Simon Bernard did not see Forts Monroe and Adams finished. In 1831 he returned to his homeland, where he served as inspector general of engineers and minister of war and where, shortly before his death, he was made a baron. Certainly because of his citizenship, his contributions were recognized more in France than in the United States. Nonetheless, his skill, vision, and particularly his largeness of conception made a profound impact on the American system of defense, and his example set a high standard for emulation by American engineers.

19. Vauban, *The New Method of Fortification as Practiced by Monsieur Vauban*, p. 22.

20. Wright, *Development of the Bastioned System*, p. 146. See also Delafield, *Report of Colonel R. Delafield*, pp. 181–82. The concept of counterscarp galleries belongs to the school of Mézières (ca. 1750).

21. Early designs for Fort Barrancas were more complex in plan than the one which was built. One proposal had a trapezoidal polygon of fortification with demibastions on the wide side overlooking Fort San Carlos. Drawing, Record Group 77, Drawer 79, Sheet 2, Cartographic Branch, National Archives, Washington, D.C.

22. For a plan of Fort Livingston, see Record Group 77, Drawer 90, Sheet 17, Cartographic Branch, National Archives, Washington, D.C.

23. "Report on the Defence of Charleston Harbor" (1827), Record Group 77, Corps of Engineers, Entry 223, p. 2, National Archives, Washington, D.C.

24. *Ibid.*, p. 5.

25. There are many documents relating to this controversy over permanent fortifications. Representing the negative side were U.S., Congress, Senate, Lewis Cass, *Report on Fortifications*, 24th Cong., 1st sess., 1836, Senate Doc. No. 293; and U.S., Congress, House, Edmund P. Gaines, *Memorial*, 26th Cong., 1st sess., 1839, House Doc. No. 206, pp. 118–43. Typical of the proponents of the system originally conceived by the board of engineers were Joseph G. Totten, *Report on the Defence of the Atlantic Frontier, from Passamaquoddy to the Sabine*, 26th Cong., 1st sess., 1840, House Doc. No. 206, pp. 5–117; and John G. Barnard, *The Dangers and Defences of New York Addressed to the Hon. J. B. Floyd, Secretary of War*.

26. Totten, *Report on the Defence of the Atlantic Frontier*, p. 66.

27. This was also the site of an earlier Fort Richmond, a semicircular sandstone castle which was constructed during the War of 1812.

Foreman, Grant. *Advancing the Frontier, 1830–1860.* Norman, Okla., 1933.

———, and Carolyn Thomas Foreman. *Fort Gibson: A Brief History.* Norman, Okla., n.d.

Fort Pitt Society. *Fort Duquesne and Fort Pitt: Early Names of Pittsburgh Streets.* Pittsburgh, 1947.

Foster, James Monroe, Jr. "Fort Bascom, New Mexico," *New Mexico Historical Review,* 35 (1960), 30–62.

Frazer, Robert W. *Forts of the West: Military Forts and Presidios and Posts Commonly Called Forts West of the Mississippi River to 1898.* Norman, Okla., 1956.

Frazier, Ida Hedrick. *Fort Recovery: An Historical Sketch Depicting Its Role in the History of the Old Northwest.* Columbus, Ohio, 1948.

Gallaher, Ruth A. *Fort Des Moines in Iowa History.* Iowa City, Iowa, 1919.

Giffen, Helen S. "Fort Miller and Millerton: Memories of the Southern Mines," *Historical Society of Southern California Quarterly,* 21 (1939), 5–16.

Gilbert, B. F. *San Francisco Harbor Defense during the Civil War.* San Francisco, 1954.

Goplen, Arnold O. "Fort Abraham Lincoln, a Typical Frontier Military Post," *North Dakota History,* 13 (1946), 176–221.

Graham, Louis E. "Fort McIntosh," *Western Pennsylvania Historical Magazine,* 15 (1932), 93–119.

Graham, Roy Eugene. "Federal Fort Architecture in Texas during the Nineteenth Century," *Southwestern Historical Quarterly,* 74 (1970), 165–88.

Grange, Roger T., Jr. "Fort Robinson, Outpost on the Plains," *Nebraska History,* 39 (1958), 192–240.

Grant, Bruce. *American Forts Yesterday and Today.* New York, 1965.

Griffin, John W. "An Archeologist at Fort Gadsden," *Florida Historical Quarterly,* 28 (1950), 254–61.

Griswold, Bert J., ed. *The Pictorial History of Fort Wayne, Indiana.* Chicago, 1917.

Hafen, LeRoy R. "Fort Jackson and the Early Trade of the South Platte," *Colorado Magazine,* 5 (1928), 9–17.

———, and Francis Marion Young. *Fort Laramie and the Pageant of the West, 1834–1890.* Glendale, Calif., 1938.

Hagen, Olaf T. "Platte Bridge Station and Fort Casper," *Annals of Wyoming,* 27 (1955), 3–17.

Haley, J. Evetts. *Fort Concho and the Texas Frontier.* San Angelo, Tex., 1952.

Hamilton, Edward P. *Fort Ticonderoga: Key to a Continent.* Boston, 1964.

Hammond, John Martin. *Quaint and Historic Forts of North America.* Philadelphia, 1915.

Hansen, Marcus L. *Old Fort Snelling.* Iowa City, 1917.

Harrington, J. C., Albert C. Manucy, and John M. Goggin. "Archeological Excavations in the Courtyard of Castillo de San Marcos, St. Augustine, Florida," *Florida Historical Quarterly,* 34 (1955), 101–41.

Hart, Herbert M. *Old Forts of the Far West.* Seattle, 1965.

———. *Old Forts of the Northwest.* Seattle, 1963.

———. *Old Forts of the Southwest.* Seattle, 1964.

———. *Pioneer Forts of the West.* Seattle, 1967.

Hieb, David L. *Fort Laramie National Monument, Wyoming.* National Park Service Historical Handbook Series, No. 20. Washington, D.C., 1954.

Historic American Buildings Survey. Collection of data sheets, photographs and drawings in Library of Congress, Washington, D.C.

Holcombe, Return I. "Fort Snelling," *American Historical Magazine,* 1 (1906), 110–33.

Holden, William Curry. "Frontier Defense, 1846–1860," *West Texas Historical Association Year Book,* 6 (1930), 35–64.

Holmes, Jack D. L. "Notes on the Spanish Fort San Esteban de Tombecbe," *Alabama Review,* 18 (1965), 281–90.

Holmes, Louis A. *Fort McPherson, Nebraska, Fort Cottonwood, N.T.: Guardian of the Tracks and Trails.* Lincoln, Nebr., 1963.

Holt, John R. *Historic Fort Snelling.* Fort Snelling, Minn., 1938.

Hultzén, Claud H., Sr. *The Story of an Ancient Gateway to the West: Old Fort Niagara.* Buffalo, N.Y., 1939.

Hunt, Elvid. *History of Fort Leavenworth, 1827–1927.* 2d ed. Fort Leavenworth, Kans., 1937.

Hunter, William A. *Forts on the Pennsylvania Frontier, 1753–1758.* Harrisburg, Pa., 1960.

Hussey, John A. *The History of Fort Vancouver and Its Physical Structure.* Portland, Ore., 1957.

Jacksonville Historical Society. *Papers.* Jacksonville, Fla., 1960.

James, Alfred Procter, and Charles Morse Stotz. *Drums in the Forest.* Pittsburgh, Pa., 1958.

Jenkins, William H. "Alabama Forts, 1700–1838," *Alabama Review,* 12 (1959), 163–79.

Jewett, Henry C. "History of the Corps of Engineers to 1914," *Military Engineer,* 14 (1922), 304–6.

Johnson, John. *The Defenses of Charleston Harbor.* Charleston, S.C., 1890.

———. "The Old Fort at Dorchester, S.C.," *South Carolina Historical and Genealogical Magazine,* 6 (1905), 127–29.

Johnson, Richard W. "Fort Snelling from Its Foundation to the Present Time," *Minnesota Historical Collections,* 8 (1898), 427–48.

Johnson, Sally A. "Fort Atkinson at Council Bluffs," *Nebraska History,* 38 (1957); 229–36.

———. "The Sixth's Elysian Fields: Fort Atkinson on the Council Bluffs," *Nebraska History,* 40 (1959), 1–38.

Jones, Robert Ralston. *Fort Washington at Cincinnati, Ohio.* Cincinnati, 1902.

Kenny, Judith Keyes. "The Founding of Camp Watson," *Oregon Historical Quarterly*, 58 (1957), 5–16.

Knight, Oliver. *Fort Worth: Outpost on the Trinity*. Norman, Okla., 1953.

Lackey, Vinson. *The Forts of Oklahoma*. Tulsa, Okla., 1963.

Lapham, Samuel, Jr. "Notes on Granville Bastion (1704)," *South Carolina Historical and Genealogical Magazine*, 26 (1925), 221–27.

Lattimore, Ralston B. *Fort Pulaski National Monument, Georgia*. National Park Service Historical Handbook Series, No. 18. Washington, D.C., 1954.

Ledyard, Edgar M. "American Posts," *The Utah Historical Quarterly*, 1 (1928), 56–64, 86–96, 114–127; 2 (1929), 25–30, 55–64, 90–96, 127–28; 3 (1930), 27–32, 59–64, 90–96; 5 (1932), 65–80, 113–28, 161–76; 6 (1933), 29–48, 64–80.

Lessem, Harold I., and George C. Mackenzie. *Fort McHenry National Monument and Historic Shrine, Maryland*. National Park Service Historical Handbook Series, No. 5. Washington, D.C., 1954.

Lorant, Stefan. *The New World: The First Pictures of America*. Rev. ed. New York, 1965.

McClellan, S. Grove. "Old Fort Niagara," *American Heritage*, 4 (1953), 32–41.

McClure, Stanley W. *The Defenses of Washington, 1861–1865*. Mimeographed. National Park Service, Washington, D.C., 1961.

Mahan, Bruce E. "Old Fort Atkinson," *Palimpsest*, 2 (1921), 333–50.

———. "Old Fort Crawford," *Palimpsest*, 42 (1961), 449–512.

———. *Old Fort Crawford and the Frontier*. Iowa City, 1926.

Mantor, Lyle E. "Fort Kearney and the Westward Movement," *Nebraska History*, 29 (1948), 175–207.

Manucy, Albert C. *The Building of the Castillo de San Marcos*. National Park Service Interpretive Series, History No. 1. Washington, D.C., 1961.

———. *The Fort at Frederica*. Notes in Anthropology, Vol. 5. Tallahassee, Fla., 1962.

———, ed. *The History of Castillo de San Marcos and Fort Matanzas from Contemporary Narratives and Letters*. Washington, D.C., 1955.

Marcum, R. T. "Fort Brown, Texas: The History of a Border Post," Ph.D. diss., Texas Technological College, 1964.

Marks, Laurence H. "Fort Mims: A Challenge," *Alabama Review*, 18 (1965), 275–80.

Mason, Philip P. *Detroit, Fort Lernoult, and the American Revolution*. Detroit, 1964.

Mattes, Merrill J. "Fort Laramie, Guardian of the Oregon Trail: A Commemorative Essay," *Annals of Wyoming*, 17 (1945), 3–20.

———. "Fort Mitchell, Scotts Bluff, Nebraska Territory," *Nebraska History*, 33 (1952), 1–34.

Mattison, Ray H. "Fort Rice—North Dakota's First Missouri River Military Post," *North Dakota History*, 20 (1953), 87–108.

Miller, Francis Trevelyan, ed. *The Photographic History of the Civil War*. 10 vols. New York, 1911.

Millis, Wade. "Fort Wayne, Detroit," *Michigan History Magazine*, 20 (1939), 21–29.

Milner, P. M. "Fort Macomb," *Publications of the Louisiana Historical Society*, 7 (1913–14), 143–52.

Mokler, Alfred James. *Fort Casper (Platte Bridge Station)*. Casper, Wyo., 1939.

Montgomery, Mrs. F. C. "Ft. Wallace and Its Relation to the Frontier," *Collections of the Kansas Historical Society*, 17 (1928), 189–283.

Montgomery, Thomas Lynch. *Frontier Forts of Pennsylvania*. 2d ed. 2 vols. Harrisburg, Pa., 1916.

Mullaly, Franklin R. "Fort McHenry, 1814: The Battle of Baltimore," *Maryland Historical Magazine*, 54 (1959), 61–103.

Mumey, Nolie. *Old Forts and Trading Posts of the West*. Denver, Colo., 1956.

Murray, Robert A. *Military Posts in the Powder River Country of Wyoming, 1865–1894*. Lincoln, Nebr., 1968.

Myers, Lee. "Fort Webster on the Mimbres River," *New Mexico Historical Review*, 41 (1966), 47–57.

Nadeau, Remi. *Fort Laramie and the Sioux Indians*. Englewood Cliffs. N.J., 1967.

Nankivell, John H. "Fort Crawford, Colorado, 1880–1890," *Colorado Magazine*, 2 (1934), 54–64.

———. "Fort Garland, Colorado," *Colorado Magazine*, 16 (1939), 13–28.

Nelson, Lee H. *An Architectural Study of Fort McHenry*. Mimeographed. Department of the Interior, National Park Service, Philadelphia, 1961.

Nye, Wilbur Sturtevant. *Carbine and Lance: The Story of Old Fort Sill*. Norman Okla., 1937.

"The Old Fort at Dorchester, S. C.," *South Carolina Historical and Genealogical Magazine*, 6 (1905), 127–30.

Oneal, Ben G. "The Beginnings of Fort Belknap," *Southwestern Historical Quarterly*, 61 (1958), 508–21.

Osterhout, George H. "The Sites of the French and Spanish Forts in Port Royal Sound," *Transactions of the Huguenot Society of South Carolina*, 41 (1936), 22–36.

Pell, S. H. P. *Fort Ticonderoga: A Short History Compiled from Contemporary Sources*. Ticonderoga, N.Y., 1966.

Peterson, Eugene T. *Michilimackinac: Its History and Restoration*. Mackinac Island, Mich., 1963.

Pfanner, Robert. "The Genesis of Fort Logan," *Colorado Magazine*, 19 (1942), 43–50.

Glossary

ABATIS. A row of obstructions made up of closely spaced, felled trees with branches trimmed to points and interlaced.

ADVANCED WORK. Any work of fortification located outside the glacis yet within musketry range.

ANGLE OF DEFENSE. The angle formed by a line of defense and a flank.

BANQUETTE. A continuous step or ledge at the base of a parapet on which defenders stood to fire over the top of the wall (fig. 113).

BARTIZAN. A projecting cylindrical form, usually located high upon a corner of a structure, which served as a watch station.

BASTION. A projection in the enceinte, made up of two faces and two flanks, which enabled the garrison to defend the ground adjacent to the enceinte.

BERM. A narrow, level space between the exterior slope and the scarp which functioned to prevent earth of the rampart from sliding into the ditch.

BLOCKHOUSE. A small fortified building used as a place of retreat or as a flanking device in forts. It was generally constructed from logs, although other materials, such as earth and stone, were commonly used in conjunction with wood.

BODY OF THE PLACE. The main enclosing fortifications from which the major defensive activities occurred.

BOMBARD. An early artillery piece that threw stones or other projectiles.

BOMBPROOF. A structure designed to provide security against artillery fire.

BULWARK. Circular works of defense surrounded by walls or ramparts.

CAPITAL OF THE BASTION. An imaginary line connecting the point of the bastion and the point of the corresponding angle of the polygon of fortification (fig. 114).

CAPONNIER. An architectural form extending from the main body of the place for the purpose of providing flanking fire. Often it served as a passage from one work to another, such as from a curtain to a ravelin. Many times it was completely enclosed and was provided with rifle ports, but other times it consisted of six- or seven-foot parapets with banquettes on the interior and glacis sloping down to the ditch on the outside of the passage.

CASEMATE. A bombproof enclosure, generally located under the rampart, for housing cannons which fired through embrasures in the scarp. Casemates were also used as quarters, magazines, and the like. In permanent fortifications they were vaulted, but in impermanent works they sometimes had trabeated structures. During the nineteenth century, rows of casemates often appeared in tiers in seacoast defenses.

CASTLE. In the medieval period, a fortified building or group of buildings. In nineteenth-century America the term also denoted a type of seacoast fortification resembling the form of a medieval shell-keep.

CAVALIER. In fortification, a raised work where artillery was placed to command the surrounding works or country. It was sometimes placed on the terreplein of a bastion or curtain.

CHEMIN DE RONDE. A narrow passage or berm located inside the scarp, at the base of the exterior slope of a rampart. The level of the path was below the top of the scarp and was thereby protected on the side facing the country. Functionally, it was used by officers to make their rounds, and it served as a place for defense against attempts at escalade.

CHEVAUX-DE-FRISE. Obstacles, mostly used in field fortification, consisting of wooden shafts from which pointed staves projected radially.

CIRCUMVALLATION. An enclosing wall, rampart, or trench, or any system of these. The term may also refer to a belt of field works.

CITADEL. A small but strong fort within, or situated to form a part of, a larger fortification or fortress. Usually located to dominate the area and other works surrounding it, it functioned as a place of refuge from which defense could be prolonged after the main works fell.

CORDON. The coping, or top course, of the scarp, normally designed to protect the wall from weathering. In plan, the line created by the cordon was termed the "magistral line" (fig. 113).

COUNTERFIRE ROOM. See COUNTERSCARP GALLERY.

COUNTERFORT. Interior buttress used to strengthen revetment walls.

COUNTERGUARD. A work made up of two faces forming a salient angle and placed before bastions or ravelins, but separated from them, to protect their faces from cannon fire.

COUNTERMINE. Underground galleries excavated by defenders for the purpose of intercepting the mines of besieging forces and destroying their works.

COUNTERSCARP. The exterior side of the ditch—the side away from the body of the place (fig. 113).

COUNTERSCARP GALLERY. A work located behind the counterscarp from which the ditch could be defended with reverse fire.

COVERED WAY. A road around a fortification between the ditch and the glacis. It was protected from enemy fire by a parapet, at the foot of which was generally a banquette enabling the coverage of the glacis with musketry. In addition to its function as an outer line of defense, it served as a place for sorties to assemble.

COVERFACE. See COUNTERGUARD.

CUNETTE. A furrow located in the bottom of a dry ditch for the purpose of drainage.

CURTAIN. A section of a bastioned fortification that lies between two bastions (fig. 112).

CURTAIN ANGLE. In plan, the angle formed between the curtain and the flank (fig. 114).

DEFENSIVE BARRACKS. Fortified quarters usually designed to serve as a citadel within a fort.

DEMIBASTION. A bastion with only one face and one flank.

DEMIGORGE. A segment lying on the prolongation of the line of the curtain and defined by the point of the curtain angle and the intersection of the line of the curtain with the capital of the bastion (fig. 114).

DEMILUNE. See HALF MOON and RAVELIN.

DETACHED BASTION. A bastion that was separated, by a space, from the main body of the place.

DETACHED SCARP. A scarp wall separated from the rampart by a *chemin de ronde*. It was also called a "Carnot wall."

DETACHED WORK. In general, a work which is beyond the range of musketry from the body of the place yet functionally related to its defense.

DITCH. A wide, deep trench around a defensive work, the material from the excavation of which was used to form the ramparts (fig. 113). When filled with water, it was termed a *moat* or *wet ditch*, otherwise it was called a *dry ditch*.

EMBRASURE. An opening in a wall or parapet through which cannons were fired. The sides, generally splayed outward, were termed *cheeks*; the bottom was called the *sole*; the narrow part of the opening, the *throat*; and the widening, the *splay*.

EMPTY BASTION. A bastion in which the interior space is much lower than the rampart—or the top of the wall if there is no rampart.

EN BARBETTE. An arrangement for cannons in which they were mounted on high platforms or carriages so that they fired over a parapet instead of through embrasures.

ENCEINTE. The works of fortification—walls, ramparts, and parapets—that enclose a castle, fort, or fortress (fig. 112).

EN CRÉMAILLÈRE. An angle or series of angles forming an orderly zig-zag line.

ENFILADE FIRE. Fire directed along the length of a ditch, parapet, wall, or the like.

ESCARP. See SCARP.

EXTERIOR DITCH. The outermost ditch between an exterior front and the covered way.

EXTERIOR FRONT. A front of fortification separated from the main body of the place by a ditch.

EXTERIOR SLOPE. A steep earth incline on the exterior side of a rampart which connects the superior slope with the ground, scarp, or berm (fig. 113).

FACE. In some five-sided seacoast forts, the designation for the sides of the enceinte which form a salient directed toward the main passage (fig. 115).

FACE OF THE BASTION. The section of any bastion between the flanked angle and the shoulder angle. In a regular bastion it was one of the two sides of the bastion which formed a salient angle pointing outwards and which was situated on the lines of defense (fig. 112).

FASCINE. A long bundle of sticks bound together for use in revetments, in stabilizing earthworks, in filling ditches, and so on.

FIELD FORTIFICATION. The art of constructing impermanent works intended for occupation only for a short time or during a campaign.

FLANK OF THE BASTION. The section of the bastion lying between the face and the curtain from which the ditch in front of the adjacent curtain and the flank and face of the opposite bastion were defended (fig. 112).

FLANK WALL. In some five-sided works, a term designating the section of the enceinte between a face and the gorge (fig. 115).

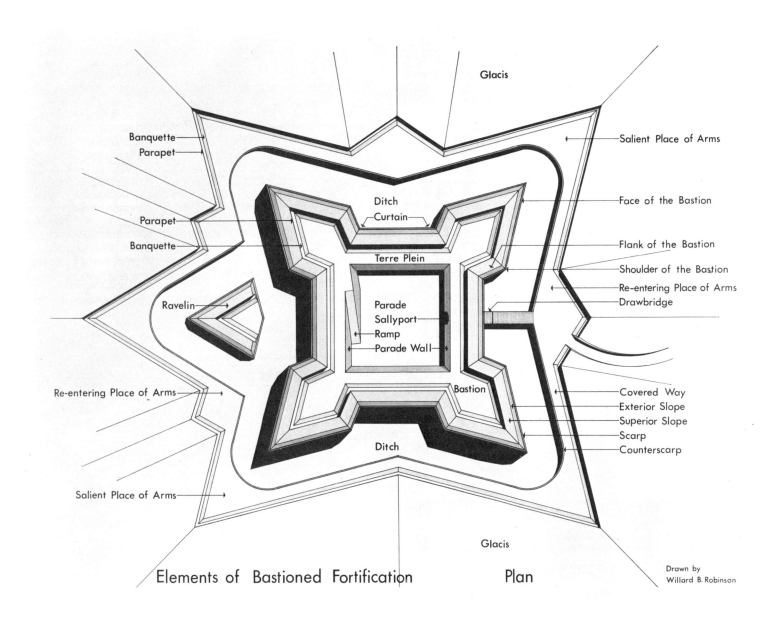

Banquette
Parapet

Parapet

Banquette

Ravelin

Re-entering Place of Arms

Salient Place of Arms

Glacis

Ditch
Curtain

Terre Plein

Parade
Sallyport
Ramp
Parade Wall

Bastion

Ditch

Glacis

Salient Place of Arms

Face of the Bastion

Flank of the Bastion

Shoulder of the Bastion

Re-entering Place of Arms
Drawbridge

Covered Way
Exterior Slope
Superior Slope
Scarp
Counterscarp

Elements of Bastioned Fortification   Plan

Drawn by
Willard B. Robinson

112. Elements of bastioned fortification.

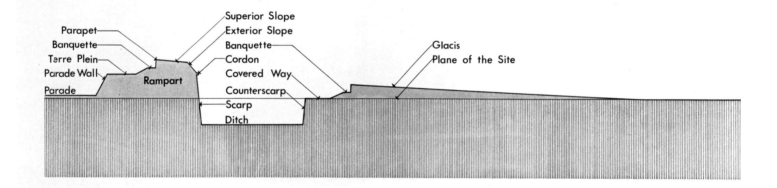

Parapet
Banquette
Terre Plein
Parade Wall
Parade

Rampart

Superior Slope
Exterior Slope
Banquette
Cordon
Covered Way
Counterscarp
Scarp
Ditch

Glacis
Plane of the Site

Profile of a Typical Bastioned or Polygonal Fortification

Drawn by
Willard B. Robinson

113. Profile of a typical bastioned or polygonal fortification.

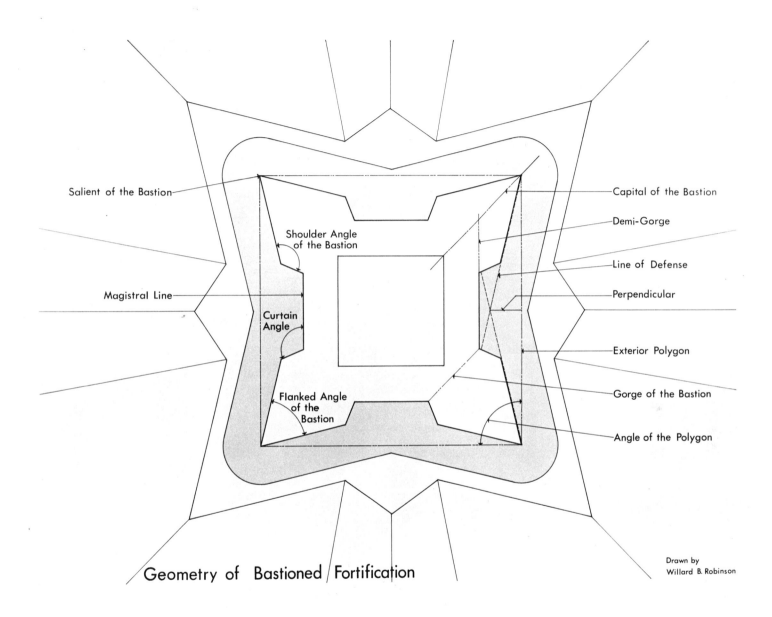

Salient of the Bastion

Shoulder Angle
of the Bastion

Magistral Line

Curtain
Angle

Flanked Angle
of the
Bastion

Capital of the Bastion

Demi-Gorge

Line of Defense

Perpendicular

Exterior Polygon

Gorge of the Bastion

Angle of the Polygon

Geometry of Bastioned Fortification

Drawn by
Willard B. Robinson

114. Geometry of bastioned fortification.

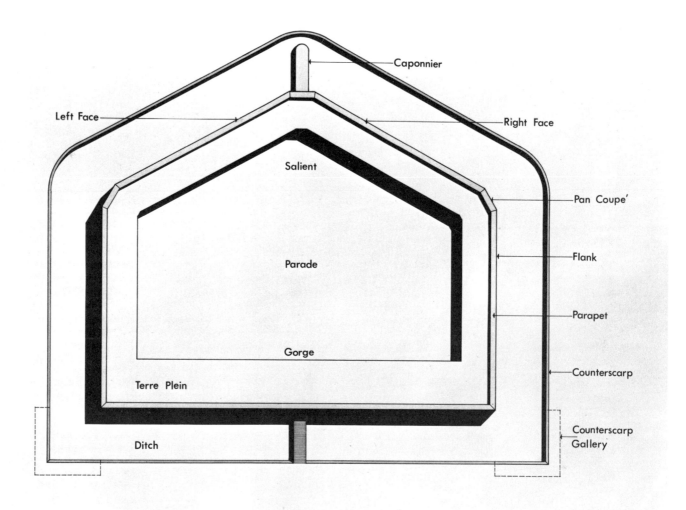

Caponnier

Left Face

Right Face

Salient

Pan Coupe'

Parade

Flank

Parapet

Gorge

Counterscarp

Terre Plein

Counterscarp Gallery

Ditch

Elements of Polygonal Casemated Fortification   Plan

115. Elements of polygonal casemated fortification.

FLANKED ANGLE. The angle formed by two faces of a bastion or ravelin (fig. 114). It is also called the "salient," the "point of the bastion," or the "point of the ravelin."

FLANKER. A projecting work from which the ground in front of adjacent walls could be defended.

FLAT BASTION. A bastion was called "flat" when it was located away from an angle in the polygon of fortification, that is, when the curtains on both sides formed a continuous, straight line. Technically stated, the gorge of a flat bastion was parallel with the demigorge.

FORT. A work established for the defense of a land or maritime frontier, of an approach to a town, or of a pass or river. Although the term originally denoted a small fortification garrisoned by troops, in North America it was used to designate virtually any establishment—civil or military—associated with protection from adversaries, regardless of whether any actual fortifications were included.

FORTIFICATION. The art of building works for defense or attack which, through their form and construction, enabled their occupants to resist assaults by superior forces for a considerable length of time.

FORTRESS. A town or city enclosed by fortifications.

FRAISE. A row of palisades planted horizontally or obliquely in the ground at the edge of a ditch or other earthwork.

FRONT OF A FORTIFICATION. The works—flanks, faces, curtains, and so on—associated with a single side of the polygon of fortification. Thus, one front of a bastioned fort consisted of two half bastions, a curtain, and related outworks.

FULL BASTION. A bastion in which the interior area is level with the terre-plein of the rampart.

GABION. A hollow cylindrical wickerwork. Open at both ends, it was set into place and then filled with earth to form parapets, embrasure cheeks, and the like, in field works.

GATE. A main entrance in the enceinte of a castle, fort, or fortress.

GLACIS. A broad, gently sloped earthwork built up outside the covered way (figs. 112, 113). At the covered way it terminated against a parapet, and in the direction of the field it sloped downward until it generally blended into the natural level of the ground.

GORGE. In a bastion, the interval or space between the two curtain angles (fig. 114). In other works that were open at the rear it denoted the opening. In some five-sided forts the designation applied to the rear section of the enceinte.

GORGE ANGLE. The angle formed by the gorge curtain and flank curtain in a five-sided fort with a polygon of fortification that was two-thirds of a hexagon (fig. 115).

GORGE BASTION. A bastion located at the angle formed by the meeting of flank and gorge walls in a five-sided polygon of fortification that was two-thirds of a hexagon. See also SALIENT BASTION and SHOULDER BASTION.

GUARDHOUSE. The headquarters for the daily guard; also a structure containing a guardroom for prisoners.

GUARDROOM. A space near the entrance of a fort where guards were stationed. Also, a room for prisoners.

HALF BASTION. See DEMIBASTION.

HALF MOON. In early Renaissance fortifications, an outwork consisting of two faces and a crescent-shaped gorge. In French, it was termed a *demilune*. Considered weak by engineers, half moons were replaced by ravelins in later works.

HORNWORK. A work made up of a bastioned front—two half bastions and a curtain—and two long sides termed *branches*. It functioned to enclose an area adjacent to, but not contained within, a fort or fortress.

ICHNOGRAPHY. The representation, with plans or horizontal sections, of the horizontal geometrical characteristics of a form.

INTERIOR SLOPE. The inner side of a parapet, generally connecting the superior slope with the banquette.

INTRENCHED CAMP. A fortified army camp located outside a fort or fortress but within the range of its guns.

INTRENCHMENT. A fieldwork comprised of a ditch and an earth parapet.

LAND FRONT. A front of a fortification designed to defend against a land-based attack.

LINE OF DEFENSE. The line extending from an angle in the exterior polygon of fortification, or flanked angle of the bastion, to the opposite flank (fig. 114). It determined the position of the face of the bastion relative to the flank which would defend it.

LISTENING GALLERY. A tunnel formed with walls and vaulting of masonry extending under various parts of earthworks from within which the sounds of enemy miners' tools could be detected.

LOOPHOLE. A small opening in a wall or stockade through which small arms were fired.

LUNETTE. A work with two faces and two parallel flanks generally used as an advanced fortification. The term also sometimes denotes a work used on the side of a ravelin.

MAGAZINE. A place for the storage of gunpowder, arms, provisions, or goods.

MAGISTRAL LINE. The base line from which the various parts of a fortification were traced.

MASK. A work concealing or protecting another work.

MERLON. The solid feature in a battlemented parapet (wall with evenly-spaced notches) forming the openings (crenels) through which archers defended the wall.

MINE. In fortification, a subterranean tunnel excavated by besiegers under a fortification for the purpose of destroying a section of the work by explosives or other means.

MOAT.   See DITCH.

MOLE.   In architecture, a foundation work raised in water with stones to form a platform for a building or fortification.

ORILLON.   A work placed at the shoulder of a bastion. It functioned as a cover for a retired flank and was used mostly in early bastioned European fortifications. There is no known case of its use in America.

ORTHOGRAPHY.   The true representation, with elevations, sections, or profiles, of the vertical characteristics of a work.

OUTWORK.   A work inside the glacis but outside the body of the place.

PALISADE.   A high fence, for defensive enclosure, made of poles or palings planted in the ground from six to nine inches apart.

PARADE.   An area, usually centrally located, where troops were assembled for drill and inspection (fig. 112).

PARADE FACE.   The wall or side of an enceinte next to the parade ground.

PARAPET.   In fortification, a work of earth or masonry forming a protective wall over which defenders fired their weapons. In buildings, a perimeter wall extending above a roof or platform.

PENTAGONAL, BASTIONED FORT.   A bastioned work developed on a polygon of fortification in the form of a pentagon.

PICKET.   A pointed pole planted vertically in the ground.

PLANE OF THE SITE.   The natural level or incline of the ground on which a fortification is constructed (fig. 113).

POLYGON OF FORTIFICATION.   A plane geometric form on which fronts of fortification were traced.

PORTCULLIS.   A sliding timber or iron grate which was suspended over a gateway and which was raised or lowered in vertical channels to close the entrance.

POSTERN.   A passage leading from the interior of a fortification to the ditch.

PROFILE.   The outline of a vertical section of work.

RAMPART.   A mass of earth formed with material excavated from the ditch to protect the enclosed area from artillery fire and to elevate defenders to a commanding position overlooking the approaches to a form or fortress (fig. 113).

RAVELIN.   A work consisting of two faces forming a salient angle which was closed at the gorge (fig. 112). Ravelins were separated from the main body of the place by ditches and functioned to protect curtains.

REDAN.   A work made up of two faces forming a salient angle open at the back.

REDOUBT.   An enclosed fortification without bastions.

REENTERING OR REENTRANT ANGLE.   An angle pointing toward the interior of a fortification.

REENTERING PLACE OF ARMS.   A space along the covered way formed outside the reentering angle of the counterscarp by providing a salient in the parapet (fig. 112). Its function was to provide space for forming sorties and a means for flanking defense of the glacis.

RETIRED FLANK.   A flank positioned inside the line of a straight flank so the shoulder of the bastion helped to protect it.

REVERSE FIRE.   Fire directed over the near side of a work to the interior of the opposite side or to the rear of a line.

REVETMENT.   The facing of the sides of a ditch or parapet.

SALIENT.   An angular work which projects outward from the interior.

SALIENT ANGLE.   An angle pointing outward.

SALIENT BASTION.   A bastion located at the salient created by the meeting of two face walls. See also GORGE BASTION and SHOULDER BASTION.

SALIENT PLACE OF ARMS.   A space along the covered way formed by rounding the trace of the counterscarp opposite the flanked angle of the bastion (fig. 112).

SALLYPORT.   A passage, either open or covered, from the covered way to the country; or a passage under the rampart, usually vaulted, from the interior of a fort to the exterior, primarily to provide for sorties.

SAP.   A trench and parapet constructed by besiegers to protect their approaches toward a fortification.

SCARP.   The interior side of the ditch (fig. 113). It was also sometimes termed *escarp*.

SECTOR OF FIRE.   The horizontal angular range of a cannon traversing on a platform.

SHOULDER ANGLE.   The interior angle formed by the meeting of a flank and a face of a bastion or other work (fig. 114).

SHOULDER BASTION.   A bastion located at the intersection of face and flank walls of a fort. The designation applied only to a particular type of five-sided seacoast defensive work that had a polygon of fortification of two-thirds of a hexagon. See also GORGE BASTION and SALIENT BASTION.

SHOULDER OF THE BASTION.   The corner formed by the intersection of the face and flank of a bastion (fig. 114).

SORTIE.   A sudden attack on besiegers by troops from a defensive work. The main objective was to destroy siege works that had been constructed by the aggressors. Also called a sally.

SQUARE, BASTIONED FORT.   A bastioned work developed on a polygon of fortification in the form of a square.

STAR FORT.   An enclosed work with a trace made up of a series of salient and reentering angles.

STOCKADE.   A defensive work—usually eight or more feet high—composed of timbers planted tightly together in the ground. Stockades were generally provided with loopholes, and since these openings were often in the upper part of the fence, banquettes or elevated walks were often necessary parts of the wall.

SUPERIOR SLOPE. The top surface of an earth parapet which slants downward toward the country, the slope of which is inclined sufficiently to allow defenders to cover all the ground outside the ditch (fig. 113).

SYSTEM OF FORTIFICATION. A formulized arrangement and proportioning of various elements of fortification, usually identified with the inventor of the system or with the country in which it was extensively used.

TALUS. A slope the function of which is to establish equilibrium in earthworks.

TENAILLE. A work which has two faces meeting to form a reentrant angle. The term also denotes a work constructed in a main ditch between bastions and in front of a curtain.

TERREPLEIN. A level space on the rampart between the parapet and the parade face (figs. 112, 113).

TOWER BASTION. A masonry bastion distinguished by its vertical characteristic. Often it was higher than the curtain ramparts.

TRACE. The outlines of the horizontal configurations of a fortification.

TRAVERSE. A parapet thrown across a covered way, a terreplein, or other location to prevent enfilade or reverse fire along a work.

TRAVERSE CIRCLE. The circle on which the wheels of a traversing carriage rode.

TRAVERSE PLATFORM. A platform which moved on tracks in an arc about a pintle. The gun carriage was mounted on the platform.

TROUS DE LOUP. An obstacle comprised of pits, each of which is an inverted pyramid or cone with a pointed stake projecting from the bottom.

TURRET. A small tower located at an angle of a building or fortification. Also, a revolving structure housing guns.

WATER FRONT. A front of fortification designed to defend against the passage of vessels.

Selected Bibliography

TREATISES ON THE ART OF FORTIFICATION

Bar-le-Duc, Jean Errard de. *La Fortification démonstrée et reduicte en art.* 2nd ed. Paris, 1620.

Barnard, John G. *Notes on Sea-Coast Defence: Consisting of Sea-Coast Fortification, the Fifteen-Inch Gun, and Casemate Embrasure.* New York, 1861.

Belidor, Bernard Forest de. *La Science des ingenieurs dans la conduite des travaux de fortification et d'architecture civile.* Paris, 1739.

Blondel, François. *Nouvelle manière de fortifier les places.* 2d ed. The Hague, 1686.

Bordwine, Joseph. "Bordwine's New System of Fortification," *Westminster Review*, 21 (1834), 480–84.

Bousmard, Henri Jean Baptiste de. *Essai général de fortification, et d'attaque et défense des places, dans lequel ces deux sciences sont expliquées et mises l'une par l'autre à la portée de tout le mode.* 2d ed. Paris, 1814.

Cambray, Chevalier de. *Manière de fortifier de Mr. de Vauban.* Amsterdam, 1689.

Carnot, Lazare Nicolas Marguerite. *De la Défense des places fortes, ouvrage composé par ordre de Sa Majesté imperiale et royale pour l'instruction de élèves du Corps du génie.* 3d ed. Paris, 1812.

Clairac, de la Mamie. *L'Ingenieur de campagne; or, Field Engineer.* Philadelphia, 1776.

Coehoorn, Minno. *The New Method of Fortification.* Trans. Thomas Savery. London, 1705.

Cormontaigne, Louis de. *Memorial pour la fortification permanente et passagere: ouvrage posthume de Cormontaigne.* 2d. ed. Paris, 1824.

Dufour, Guillaume Henri. *De la Fortification permanente.* Geneva, 1822.

Fer, Nicolas de. *Les Forces de l'Europe ou description des principales villes avec leurs fortifications.* Paris, 1690–95.

———. *Introduction a la fortification.* Paris, 1724.

Fournier, George. *Traité des fortifications, ou architecture militaire, tirée des places les plus estimées de ce temps, pour leurs fortifications, devisé en deux parties.* Paris, 1647.

The Gentleman's Compleat Military Dictionary. . . . 18th ed. Boston, 1759.

Holden, Edward S. *Notes on the Bastion System of Fortification: Its Defects and Their Remedies.* New York, 1872.

Ive, Paul. *The Practice of Fortification.* London, 1589.

James, Charles. *A New and Enlarged Military Dictionary . . . of the Different Systems of Fortification. . . .* London, 1805.

Le Blonde, M. *Éléments de fortification.* 5th ed. Paris, 1764.

———. *Traité de la defense des places, avec . . . une dictionaire des terms de l'artillerie, de fortification, de l'attaque & de la defense des places.* 2d ed. Paris, 1762.

Lendy, A. F. *Treatise on Fortification; or, Lectures Delivered to Officers Reading for the Staff.* London, 1862.

Léry, Gaspard-Joseph Chaussegros de. "Traité de fortification." Manuscript Group 18, Series K2, Vol. 1, Public Archives of Canada, Ottawa.

Mahan, D. H. *An Elementary Course of Military Engineering.* 2 vols. New York, 1866, 1867.

———. *An Elementary Treatise on Advanced-Guard, Out-Post, and Detachment Service of Troops, and the Manner of Posting and Handling Them in the Presence of an Enemy. . . .* New York, 1853.

Marchi, Francesco de. *Architettura militare di Francesco d' Marchi, illustrata da Luigi Marini.* Rome, 1810.

Marlois, Samuel. *Fortification ou architecture militaire.* Amsterdam, 1627.

Martini, Francesco di Giorgio. *Trattato di architettura civile e militare di Francesco di Giorgio Martini. . . .* Turin, 1841.

A Military Dictionary Explaining All Difficult Terms in Martial Discipline, Fortification, and Gunnery. London, 1708.

"Modern Fortification," *Edinburgh Review*, 102 (1855), 202–36.

Montalembert, Marc-René, Marquis de. *La Fortification perpendiculaire, ou essai sur plusieurs manières de fortifier la ligne droite, le triangle, le quarré et tous les polygônes, de quelqu'entendue qu'en soient les côtés, en donnant à leur défense une direction perpendiculaire.* 4 vols. Paris, 1776–78.

Muller, John. *Attack and Defense of Fortify'd Places.* London, 1747.

———. *The Field Engineer of M. le Chevalier de Clairac. . . .* London, 1760.

———. *A Treatise Containing the Elementary Part of Fortification, Regular and Irregular.* 2d ed. London, 1756.

Pagan, Blaise-Françoise, Comte de. *Traité des fortifications.* Paris, 1645.

Piron, F. P. J. "The Systems of Fortification Discussed and Compared," *United States Service Magazine*, 5 (1866), 34–40, 108–14, 225–34, 328–32.

Scheliha, Viktor Ernst Karl Rudolf von. *A Treatise on Coast-Defence: Based on the Experience Gained by Officers of the Corps of Engineers of the Army of the Confederate States. . . .* London, 1868.

Scott, H. L. *Military Dictionary. . . .* New York, 1864.

Straith, Hector. *Treatise on Fortification and Artillery.* 7th ed. Revised by Thomas Cook and John T. Hyde. London, 1858.

Tousard, Louis de. *American Artillerists Companion; or, Elements of Artillery. . . .* 3 vols. Philadelphia, 1809–13.

Vauban, Sébastien de Prestre de. *De l'Attaque et de la defense des places.* Ed. Pierre de Honet. 2 vols. The Hague, 1737–42.

———. *Military Engineering.* Amsterdam, 1689.

———. *The New Method of Fortification as Practiced by Monsieur Vauban . . . Made English.* Trans. Abel Swall. London, 1691.

———. "Traité de fortifications ou l'on explique les differens sistemes oui ont eté misusage jusqu'ce jour et les differens dehors qui ont eté practiquez pour eloigner les aproaches," Manuscript, United States Military Academy Library, West Point, N.Y.

———. *Traité de l'attaque des places.* Paris, 1795.

Vernon, Simon François Gay de. *Traité elementaire d'art militaire et de fortification à l'usage élèves de l'École polytechnique, et de élèves de écoles militaires.* 2 vols. Paris, 1805.

———. *A Treatise on the Science of War and Fortifications.* 2 vols. Trans. John Michael O'Connor. New York, 1817.

Viollet-le-Duc, Eugène Emmanuel. *Military Architecture.* Trans. M. MacDermott. 2d ed. Oxford, 1879.

Woodruff, J. A. *Applied Principles of Field Fortification for Line Officers.* Leavenworth, Kans., 1909.

Wright, John Womack. *The Development of the Bastioned System of Fortifications, 1500–1800.* Washington, 1946.

Manuscript Material

Archives du Ministère des Armées, Vincennes. Plan.

Archives Nationales, Section Outre-mer, Paris. Maps and plans of American fortifications.

Archivo General de Indias, Seville. Miscellaneous plans and drawings of American fortifications.

Bernard, Simon. Miscellaneous reports, Record Group 77, Entry 20, File A, Nos. 20, 22, 28, 35, National Archives, Washington, D.C.

Bibliothèque Nationale, Department des Cartes et Plans, Paris. Miscellaneous maps and plans.

British Museum, Map Room, London. Miscellaneous maps and plans.

Corps of Engineers. Buell Collection on microfilm, 3 rolls, No. M-417, National Archives, Washington, D.C.

———. "Daily Reports of Operations on Construction of Fort Morgan, Mobile Point (January, 1828–May, 1829)," Record Group 77, Entry 1266, National Archives, Washington, D.C.

———. "Journal of Operations, Fort Gaines (February, 1857–February, 1860)," Record Group 77, Entry 1267, National Archives, Washington, D.C.

———. "Letters Received at Fort Clinch, Amelia Island, Florida (October, 1865–September, 1866)," Record Group 77, Entry 1225, National Archives, Washington, D.C.

———. "Letters Sent from Fort Morgan, Mobile Point (May, 1821–March, 1822; July, 1824–April, 1828; November, 1845–June, 1850)," Record Group 77, Entry 1237, 2 vols., National Archives, East Point, Ga.

———. "Newport, Rhode Island, Engineer Office, Letters Sent and Received Relating to Fort Adams," Record Group 77, Entry 616, National Archives, Waltham, Mass.

———. "Reports of Boards of Engineers Relating to Fortifications and Defenses, 1821–34," Record Group 77, Entry 223, National Archives, Washington, D.C.

Fessenden, F. M. "The Building of Fort Phillip Kearney the Summer of 1866," E. A. Brininstool Collection, University of Texas Library, Austin.

Gadsden, James. "Report upon the Defenses of the Gulf of Mexico, November 10, 1817," Record Group 77, Office of the Chief of Engineers, Reports, July 3, 1812, to October 4, 1823, National Archives, Washington, D.C.

John Carter Brown Library, Providence, R.I. Collection of maps, drawings, and prints of early American fortifications.

National Archives, Cartographic Branch, Washington, D.C. Collection of maps, plans, drawings, and miscellaneous reports pertaining to American forts.

_____, Old Army Branch, Washington, D.C. Miscellaneous letters.

_____, Still Pictures Branch, Washington, D.C. Collection of photographs of American forts.

Public Archives of Canada, Manuscript Division, Ottawa. Miscellaneous letters and reports.

_____, Map Division, Ottawa. Collection of maps, plans, and drawings of American forts.

Public Record Office, London. Miscellaneous maps and plans.

Totten, Joseph G. Totten Papers, 7 vols, Record Group 77, Entry 147, National Archives, Washington, D.C.

William L. Clements Library, Ann Arbor, Mich. Plans and drawings of fortifications of the Revolutionary War.

Williams, Jonathan. "General Return of Fortifications from Massachusetts to Pennsylvania," 1802, Record Group 77, Drawer 245, Sheet 6, National Archives, Washington, D.C.

GOVERNMENT DOCUMENTS

American State Papers: Military Affairs. 7 vols. Washington, 1860.

Barnard, John G. "Eulogy on the Late Joseph G. Totten," *Annual Report of the Smithsonian Institution.* 39th Cong., 1st sess., 1866, House Exec. Doc. No. 102.

Belknap, W. W. *Report on Work on Fort Tompkins.* 42nd Cong., 2d sess., March 7, 1872, House Exec. Doc. No. 189.

Blair, F. P., Jr. *Permanent Fortifications and Sea-Coast Defences.* 37th Cong., 2d sess., 1862, Rept. No. 86.

Calhoun, John C. *Letter Relating to Sites of Fortifications.* 16th Cong., 2d sess., February 12, 1821, Exec. Paper No. 98.

_____. *Report on Fortifications.* 16th Cong., 1st sess., January 15, 1820, Exec. Paper No. 51.

Cameron, Simon. *Report on Fortifications.* 37th Cong., 2d sess., December 12, 1861, House Exec. Doc. No. 6.

Cass, Lewis. *Report on Fortifications.* 24th Cong., 1st sess., 1836, Senate Doc. No. 293.

Conrad, C.M. *Report of the Secretary of War.* 32d Cong., 1st sess., November 29, 1851, House Exec. Doc. No. 2.

_____. *Report on Fortifications.* 32d Cong., 1st sess., December 8, 1851, House Exec. Doc. No. 5.

_____. *Report on Forts Jefferson and Taylor.* 32d Cong., 2d sess., January 20, 1853, Senate Exec. Doc. No. 25.

Coolidge, Richard H. *Statistical Report on the Sickness and Mortality in the Army of the United States, Compiled from the Records of the Surgeon General's Office Embracing a Period of Sixteen Years, from January, 1839, to January, 1855.* 34th Cong., 1st sess., 1856, Senate Exec. Doc. No. 96.

Davis, Jefferson. *Report on Fortifications.* 34th Cong., 1st sess., March 20, 1856, House Exec. Doc. No. 64.

_____. *Report on Fortifications at Tortugas.* 33d Cong., 1st sess., May 19, 1854, Senate Exec. Doc. No. 66.

Dearborn, H. *Report on Coast Defense.* 10th Cong., 1st sess., December 8, 1807, Exec. Docs.

_____. *Report on Fortifications of Ports and Harbors.* 10th Cong., 2d sess., January 7, 1809, Exec. Repts.

_____. *Report on Seacoast Defences.* 9th Cong., 1st sess., February 18, 1806, Exec. Docs.

Delafield, R. *Report of Colonel R. Delafield, U.S. Army, on the Art of War in Europe in 1854, 1855–1856.* 36th Cong., 1st sess., 1860, Senate Exec. Doc. No. 59.

Endicott, William C., *et al. Report of the Board on Fortifications or Other Defenses Appointed by the President of the United States.* 49th Cong., 1st sess., 1886, Exec. Doc. No. 49.

Fillmore, Millard. *Statements as to Appropriations for Fortifications.* 27th Cong., 1st sess., June 22, 1841, House Doc. No. 30.

Gaines, Edmund P. *Memorial.* 26th Cong., 1st sess., December 31, 1839, House Doc. No. 206.

Halleck, H. W. *Message on National Defense.* 28th Cong., 2d sess., February 7, 1845, Senate Doc. No. 85.

Marcy, W. L. *Report of the Secretary of War.* 29th Cong., 2d sess., January 7, 1847, Senate Doc. No. 44.

Mitchell, S. L. *Report on Sea-Coast Defences.* 10th Cong., 1st sess, December 3, 1807, Exec. Docs.

_____. *Report on the Defences of New York Harbor.* 9th Cong., 1st sess., January 28, 1806, Exec. Docs.

Morton, James St. C. *Memoir on American Fortifications.* 36th Cong., 1st sess., December 27, 1859, Senate Doc. No. 2.

Paulding, J. K. *Report on Military and Naval Defences.* 26th Cong., 1st sess., January 24, 1840, Senate Doc. No. 120.

Pendleton, N. G. *Military Posts—Council Bluffs to the Pacific Ocean.* 27th Cong., 3d sess., 1843, House Rept. No. 31.

Pickering, Timothy. "Report of the Department of War, Relative to the Fortifications of the Ports and Harbors of the United States," *Annals of Congress,* 4th Cong., 2d sess., January 16, 1796, 6:2571–72.

Poinsett, J. R. *Letter from the Secretary of War to the Hon. R. M. T. Hunter, May 12, 1840.* 26th Cong., 1st sess., 1840, House Doc. No. 206.

_____. *Report on Military and Naval Defences.* 26th Cong., 1st sess., April 24, 1840, Senate Rept. No. 451.

Richardson, J. D., ed. *A Compilation of the Messages of the Presidents, 1789–1902.* 10 vols. 53d Cong., 2d sess., 1907, House Misc. Doc. No. 210, Pts. 1–10.

Stanton, Benjamin. *Report on Condition of Forts and Arsenals.* 36th Cong., 2d sess., February 18, 1861, House Rept. No. 85.

Totten, J. G. *Annual Report of Chief Military Engineer.* 27th Cong., 2d sess., November 19, 1841, House Doc. No. 2.

————. *Annual Report of the Engineer Bureau.* 37th Cong., 2d sess., December 3, 1861, Senate Doc. No. 1.

————. *Report of the Chief Engineer.* 29th Cong., 2d sess., November 10, 1846, House Exec. Doc. No. 4.

————. *Report of the Colonel of Engineers.* 32d Cong., 1st sess., 1850–51, House Exec. Doc. No. 2.

————. *Report of the Engineer Department.* 32d Cong., 2d sess., November 30, 1852, House Exec. Doc. No. 1.

————. *Report on Fortifications.* 32d Cong., 1st sess., 1851, House Exec. Doc. No. 5.

————. *Report on the Defence of the Atlantic Frontier, from Passamaquoddy to the Sabine.* 26th Cong., 1st sess., May 12, 1840, House Doc. No. 206.

U.S. Congress. *Report Relating to Fortifications.* 17th Cong., 1st sess., March 4, 1822, Exec. Paper No. 89.

U.S. Congress. House. *Military and Naval Defences.* 24th Cong., 1st sess., April 7, 1836, House Doc. No. 243.

U.S. Congress. Senate. *Indian Operations on the Plains.* 50th Cong., 1st sess., April 4, 1887, Senate Exec. Doc. No. 33.

————. *Report of the Secretaries of War and Navy Relative to the Military and Naval Defences of the Country.* 24th Cong., 1st sess., 1836, Senate Doc. No. 293.

————. *Report on Fortifications.* 32d Cong.. 1st sess., March 23, 1842, Senate Rept. No. 141.

U.S. Department of the Army. Surgeon General's Office. *Circular No. 4: A Report on Barracks and Hospitals with Descriptions of Military Posts.* Washington, 1870.

————. *Circular No. 8: A Report on the Hygiene of the United States Army with Descriptions of Military Posts.* Washington, 1875.

Primary Sources: Books and Articles

Barnard, John G. *The Dangers and Defences of New York Addressed to the Hon. J. B. Floyd, Secretary of War.* New York, 1859.

————. "The Use of Iron in Fortifications," *United States Service Magazine,* 1 (1864), 25–31.

Barres, Joseph F. W. des. *The Atlantic Neptune Published for the Use of the Royal Navy of Great Britain.* 2 vols. London, 1780, 1781.

The Battle of Fort Sumter and the First Victory of the Southern Troops, April 13th 1861. Charleston, 1861. Reprint ed., Charleston, 1961.

Beauregard, P. G. T. "The Defense of Charleston," *North American Review,* 142 (1886), 419–36.

Bellin, M. *Le Petit atlas maritime recueil de cartes et plans des quatre parties du monde,* vol. I, *Contenant l'Amerique septentrionale et les isles antilles.* N.p., 1764.

Braddock's Expedition and Battle of Lake George, 1755. Atlas of engraved maps in John Carter Brown Library, Providence.

Bry, Theodor de. *Narrative of Le Moyne, an Artist Who Accompanied the French Expedition to Florida under Laudonniere, 1564.* Boston, 1875.

Chinard, Gilbert. *George Washington as the French Knew Him: A Collection of Texts.* Princeton, 1940.

Choris, Louis. *Voyage pittoresque autour de monde....* Paris, 1822.

Christoflesturm, Leonhard. *Le Veritable Vauban se montrant au lieu du faux Vauban.* The Hague, 1708.

Collot, Victor. *A Journey in North America....* Paris, 1826.

Crimmins, M. L., ed. "Colonel J. K. F. Mansfield's Report of the Inspection of the Department of Texas in 1856," *Southwestern Historical Quarterly,* 42 (1938–39), 122–48, 215–57, 351–87.

Cullum, George W. *Biographical Register of the Officers and Graduates of the U.S. Military Academy from 1802 to 1878.* 2 vols. New York, 1879.

"The Defense of Our Sea-Ports," *Harper's New Monthly Magazine,* 71 (1885), 928–37.

Department of War. "Report on Fortifications," *Niles' Register,* 20 (1821), 263–69, 285.

Dumont, M. *Memoires historiques sur la Louisiane.* 2 vols. Paris, 1753.

Flores, Alvaro. "The Visitation Made by Alvaro Flores of the Forts of Florida," *Colonial Records of Spanish Florida,* 2: 116–203.

Franklin, W. B. "National Defense," *North American Review,* 138 (1883), 594–604.

Gadsden, James. "The Defenses of the Floridas, August 1, 1818." Report reprinted in *The Florida Historical Quarterly,* 15 (1937), 242–48.

Gillmore, Q. A. *Engineer and Artillery Operations against the Defences of Charleston Harbor in 1863....* New York, 1865.

————. *Official Report to the United States Engineer Department, on the Siege and Reduction of Fort Pulaski, Georgia....* Papers on Practical Engineering, No. 8. New York, 1862.

————. *Supplementary Report to the Engineer and Artillery Operations against the Defences of Charleston Harbor in 1863.* New York, 1868.

Grant, Ulysses S. *Personal Memoirs of U. S. Grant.* 2 vols. New York, 1885–86.

Griffin, Eugene. "Our Sea-Coast Defenses," *North American Review,* 147 (1888), 64–75.

Hakluyt, Richard. *The Principal Navigations, Voyages, Traffiques and Discoveries of the English Nation.* 5 vols. London, 1810–12.

Johnston, John. *The Defense of Charleston Harbor Including Fort Sumter and the Adjacent Islands, 1863–1865*. Charleston, 1889.

Joutel, Henri. *Joutel's Journal of La Salle's Last Voyage, 1684–7*. Albany, N.Y., 1906.

Kite, Elizabeth. *Brigadier-General Louis Lebeque Duportail, Commandant of Engineers in the Continental Army, 1777–1783*. Baltimore, 1933.

La Salle, Robert Cavelier, Sieur de. *Relation of the Discoveries and Voyages of Cavelier de la Salle from 1679 to 1681*. Trans. Melville B. Anderson. Chicago, 1901.

Léry, Gaspard-Joseph Chaussegros de. "Journal, May 8th to July 2nd, 1756," *Bulletin of the Fort Ticonderoga Museum*, 6 (1942), 128–44.

———. *Inventaire des papiers de Léry conservés aux archives de la province de Quebec*. 3 vols. Ed. Pierre-Georges Roy. Quebec, 1939–40.

List of Military Posts and Stations of the United States, with Their Garrisons and Also the Stations of Troops by Companies. Washington, 1871.

Mansfield, Joseph K. F. *Mansfield on the Condition of the Western Forts, 1853–54*. Ed. Robert W. Frazer. Norman, Okla., 1963.

Marcy, R. B. *Outline Descriptions of the Posts and Stations of Troops in the Geographical Divisions and Departments of the United States*. Washington, 1872.

Margry, Pierre. *Découvertes et etablissements des français dans l'ouest et dans le sud de l'Amerique septentrionale (1614–1754)*. 6 vols. Paris, 1879–88.

Morton, James St. C. *Letter to the Hon. John B. Floyd, Presenting for His Consideration a New Plan for the Fortification of Certain Points of the Seacoast of the United States*. Washington, 1858.

———. *Memoir on American Fortification Submitted to the Hon. John B. Floyd, Secretary of War*. Washington, 1859.

Motte, Jacob Rhett. *Journey into Wilderness: An Army Surgeon's Account of Life in Camp and Field during the Creek and Seminole Wars, 1836–1838*. Ed. James F. Sunderman. Gainesville, Fla, 1963.

Outline Descriptions of the Posts in the Military Division of the Missouri Commanded by Lieutenant General P. H. Sheridan. Chicago, 1872.

Outline Descriptions of the Posts in the Military Division of the Pacific, Major-General Irwin McDowell Commanding. San Francisco, Calif., 1879.

Parkman, Francis. *The Oregon Trail: Sketches of Prairie and Rocky Mountain Life*. Ed. Charles H. J. Douglas. New York, 1916.

Peña, J. A. de la. *Derrotero de la expedición en la Provincia de los Texas. . . .* Mexico, 1722.

Pichon, Thomas. *Lettres et mémoires pour servir a l'histoire naturelle, civile et politique du Cap Breton. . . .* The Hague, 1760.

Pittman, Philip. *The Present State of the European Settlements on the Mississippi*. Reprint of 1770 ed. Cleveland, 1906.

Potter, Woodburne. *The War in Florida: Being an Exposition of Its Causes, and an Accurate History of the Campaigns of Generals Clinch, Gaines, and Scott*. Baltimore, 1836.

Poussin, Guillaume Tell. *Les Etats-Unis d'Amerique. . . .* Paris, 1874.

Ranson, Edward. "The Endicott Board of 1885–86 and the Coast Defenses," *Military Affairs* (Summer, 1967), 74–84.

Riley, Edward M., and Charles E. Hatch, Jr., eds. *James Towne in the Words of Contemporaries*. Rev. ed. National Park Service Source Book Series, No. 5. Washington D.C., 1955.

A Set of Plans and Forts in America Reduced from Actual Surveys. London, 1765.

Smalley, H. A. "A Defenseless Sea-Board," *North American Review*, 137 (1884), 233–45.

Sparks, Jared. *The Writings of George Washington: Being His Correspondence, Addresses, Messages, and Other Papers, Official and Private*. 12 vols. Boston, 1833–39.

Stokes, I. N. Phelps. *The Iconography of Manhattan Island, 1498–1909*. New York, 1915.

———, and Daniel C. Haskell. *American Historical Prints: Early Views of American Cities, etc., from the Phelps Stokes and Other Collections*. New York, 1933.

Thwaites, Reuben Gold, and Louise Phelps Kellogg, eds. *Frontier Defense on the Upper Ohio, 1777–1778*. Draper Series, Vol. III. Madison, Wis., 1912.

Totten, Joseph Gilbert. *Report Addressed to the Hon. Jefferson Davis, Secretary of War, on the Effects of Firing with Heavy Ordnance from Casemate Embrasures. . . .* Papers on Practical Engineering, No. 6. Washington, 1857.

———. *Report of General J. G. Totten, Chief Engineer, on the Subject of National Defences*. Washington, 1851.

Underhill, John. *Newes from America; or A New Discoverie of New England*. London, 1638.

Van Vliet, Stewart. *Outline Descriptions of Posts and Stations of Troops in the Military Division of the Atlantic, Commanded by Major General George G. Meade*. Philadelphia, 1870.

Vinci, Leonardo da. *The Notebooks of Leonardo da Vinci*. 2 vols. Trans. Edward MacCurdy. New York, 1958.

Webster, Noah. "Letters from Noah Webster to Ezra Stiles Respecting the Fortifications in the Western Country," *Carey's American Museum*, 6 (1789), 27–30, 136–41; 7 (1790), 323–28.

Wislizenus, Adolphus. *A Journey to the Rocky Mountains in the Year 1839*. Saint Louis, 1912.

SECONDARY SOURCES: HISTORIES

Alexander, Thomas G., and Leonard J. Arrington. "The Utah Military Frontier, 1872–1912: Forts Cameron, Thornburg, and Duchesne," *Utah Historical Quarterly*, 32 (Fall, 1964), 330–54.

Arnade, Charles W. *The Siege of St. Augustine in 1702*. Gainesville, Fla., 1959.

Arthur, Robert. *History of Fort Monroe*. Fort Monroe, Va., 1930.

Ashcraft, Allan C. "Fort Brown, Texas, in 1861," *Texas Military History*, 3 (1963), 243–47.

Athearn, Robert G. *Forts of the Upper Missouri*. Englewood Cliffs, N.J., 1967.

Barnes, Frank. *Fort Sumter National Monument, South Carolina*. National Park Service Historical Handbook Series, No. 12. Rev. ed. Washington, D.C., 1962.

Barnwell, Joseph W. "Fort King George," *The South Carolina Historical and Genealogical Magazine*, 27 (1926), 189–203.

Barrett, Arrie. "Western Frontier Forts of Texas," *West Texas Historical Association Year Book*, 7 (1931), 115–39.

Barry, Richard Schriver. "Fort Macon: Its History," *North Carolina Historical Review*, 27 (1950), 163–77.

Bearss, Edwin C. "The Fall of Fort Henry, Tennessee," *West Tennessee Historical Society*, 17 (1963).

––––––. "Unconditional Surrender. The Fall of Fort Donelson," *Tennessee Historical Quarterly*, 21 (1962).

Bender, Averam B. *The March of Empire: Frontier Defense in the Southwest, 1848–1860*. Lawrence, Kans., 1952.

Bennet, Charles E. "Fort Caroline, Cradle of American Freedom," *Florida Historical Quarterly*, 35 (1956), 3–16.

––––––. *Laudonniere and Fort Caroline*. Gainesville, Fla., 1964.

Bierschwale, Margaret. *Fort McKavett, Texas: Post on the San Saba*. Salado, Tex., 1966.

Boyd, Mark F. "Events at Prospect Bluff on the Apalachicola River, 1808–1818," *Florida Historical Quarterly*, 16 (1937), 55–96.

––––––. "The Fortifications at San Marcos de Apalache," *Florida Historical Quarterly*, 15 (1936), 3–34.

Bradford, Sydney. "Fort McHenry, 1814: The Outworks in 1814," *Maryland Historical Magazine*, 54 (1959), 188–209.

Braly, Earl Burk. "Fort Belknap of the Texas Frontier," *West Texas Historical Association Year Book*, 30 (1954), 83–114.

Brandes, Ray. *Frontier Military Posts of Arizona*. Globe, Ariz., 1960.

Brice, Wallace A. *History of Fort Wayne from the Earliest Known Accounts of This Point, to the Present Period*. Fort Wayne, 1868.

Brown, Dee. *Fort Phil Kearny: An American Saga*. New York, 1962.

Calderon Qui jano, J. A. *Fortificaciones en Nueva España*. Seville, 1953.

Carter, Kate B. *Heart Throbs of the West*. Salt Lake City, Utah, 1941.

Chapman, John. "Fort Concho," *Southwest Review*, 25 (1940), 426–45.

––––––. "Fort Griffin," *Southwest Review*, 27 (1942), 426–45.

––––––. "Old Fort Richardson," *Southwest Review*, 38 (1953), 62–69.

Chatelain, Verne E. *The Defenses of Spanish Florida, 1565 to 1763*. Carnegie Institution of Washington Publication 511. Washington, D.C., 1941.

Clinton, Amy Cheney. "Historic Fort Washington," *Maryland Historical Magazine*, 32 (1937), 228–47.

Connor, Jeannette Thurber. "The Nine Old Wooden Forts of St. Augustine," *Florida Historical Quarterly*, 4 (1926), 103–11, 171–80.

Covington, James W. "Life at Fort Brooke, 1824–1836," *Florida Historical Quarterly*, 36 (1958), 319–30.

Crimmins, M. L. "Camp Cooper and Fort Griffin, Texas," *West Texas Historical Association Year Book*, 17 (1941), 32–43.

––––––. "Fort Elliott, Texas," *West Texas Historical Association Year Book*, 23 (1947), 3–12.

––––––. "Fort Fillmore," *New Mexico Historical Review*, 6 (1931), 327–33.

––––––. "Fort McKavett, Texas," *Southwestern Historical Quarterly*, 38 (July, 1934), 28–39.

––––––. "Fort Massachusetts, First United States Military Post in Colorado," *Colorado Magazine*, 14 (1937), 128–35.

Cubberly, Frederick. "Fort King," *Florida Historical Society Quarterly*, 5 (1927), 139–52.

Cullimore, Clarence C. *Old Adobes of Forgotten Fort Tejon*. Bakersfield, Calif., 1949.

Culverwell, Albert. *Stronghold in the Yakima Country: The Story of Fort Simcoe, 1856–1859*. Olympia, Wash., 1956.

Currey, J. Seymour. *The Story of Old Fort Dearborn*. Chicago, 1912.

Davis, Mrs. Elvert M. "Fort Fayette," *Western Pennsylvania Historical Magazine*, 10 (1927), 65–84.

Dunnack, Henry E. *Maine Forts*. Augusta, Maine, 1924.

Emery, B. Frank. "Fort Saginaw," *Michigan History Magazine*, 30 (1946), 476–503.

––––––. *Fort Saginaw, 1822–1823: The Story of a Forgotten Frontier Post*. Detroit, 1932.

Emmett, Chris. *Fort Union and the Winning of the Southwest*. Norman, Okla., 1965.

Everhart, William C. *Vicksburg National Military Park, Mississippi*. National Park Service Historical Handbook Series, No. 21. Washington D.C., 1954.

Favrot, H. Mortimer. "Colonial Forts of Louisiana," *Louisiana Historical Quarterly*, 26 (1946), 722–54.

Faye, Stanley. "The Arkansas Post of Louisiana: French Domination," *Louisiana Historical Quarterly*, 26 (1946), 633–721.

––––––. "British and Spanish Fortifications of Pensacola, 1721–1821," *Florida Historical Quarterly*, 20 (1941), 277–92.

––––––. "Spanish Fortifications of Pensacola, 1698–1763," *Florida Historical Quarterly*, 20 (1941), 151–68.

Fletcher, Henry T. "Old Fort Lancaster," *West Texas Scientific Society Publications*, 4 (1932), 33–44.

28. The United States was evidently one of the few countries, if not the only one, which adopted French theory so completely. By the middle of the nineteenth century most European countries followed the German system or variations of it. The complete dedication of American military engineers to the French systems was indicated by one of the early reports of a visitation committee to the United States Military Academy. After reviewing the academic program on fortification and the textbooks used, they questioned whether the original version, rather than the translation, of Gay de Vernon's treatise on military architecture should be used. Cf. *American State Papers: Military Affairs,*, vol. 3, p. 144.

However, while forms closely conformed to the French theory, details were often freely modified. Probably the most significant of these was the design of casemate embrasures. As a result of studies made by Joseph G. Totten, the throat of American embrasures was positioned near the center of the wall, not the interior face, and the size of the opening was thereby reduced; this lessened the chances of enemy cannon shots being funneled into the casemates. Offsets in the cheeks of the embrasures also reduced the chances of missiles ricocheting inward, and the use of brick instead of stone for the cheeks reduced the injury of gunners from flying splinters caused by shot striking the sides of the openings. There were many variations in the designs of embrasures, all dependent upon their precise function, location, exposure to enemy cannon fire, and the type of cannons to be mounted. In Fort Adams, Rhode Island, for example, there are over a dozen different embrasures, varying in height, width, cheek design, and provisions for cannon carriage anchors.

29. Delafield, *Report of Colonel R. Delafield*, p. 19. See fig. 100 for a section drawing of this configuration in an American work.

30. Carnot received technical military training at Mézières and served as an engineer but was a versatile contributor to many fields. His genius was expressed in scientific, political, and literary, as well as military, works. His main contribution to the art of fortification was *De la Défense des places fortes, ouvrage composé par ordre de Sa Majesté imperiale et royale pour l'instruction de élèves du Corps du génie*, a work which became a classic in Europe. Charles François Dumouriez, a French general, credited Carnot as "... le createur du nouvel art militaire en France...." *Biographie universelle*, s.v. "Carnot, Lazare Nicolas Marguerite." Although he wrote *Éloge de Vauban* (Dijon, 1784), he rejected the theories of Vauban and his followers in favor of his own and those of Montalembert that he adopted. He also served as minister of war under the Republic and as minister of the interior under Napoleon.

31. Fort Gaines was constructed on or near the site of the fort designed by Bernard which was commenced in 1819 but abandoned shortly thereafter. After many years of delay, the development of steamships finally prompted its construction. There are two widely separated main channels into Mobile Bay, one of which was defended by Fort Morgan. The channel covered by Fort Gaines was shallow and could not be navigated by sailing vessels of war—the first proposed fort was to secure the island from an enemy which might have benefited from establishing permanent quarters there. War steamers, however, had a shallower draft and could have navigated the channel next to Dauphin Island, out of range of the guns at Mobile Point.

32. Totten, *Report on the Defence of the Atlantic Frontier*, p. 41. Totten was, of course, quite familiar with the work of the esteemed French engineer. Carnot's *Defence des places* was listed among the numerous books Totten placed in the library of the United States Military Academy. The complete list of titles appears in Record Group 77, Entry 146, Vol. III, pp. 187–91, National Archives, Washington, D.C.

33. For a brief history and description of North American Martello towers, see Willard B. Robinson, "North American Martello Towers," *Journal of the Society of Architectural Historians*, 33 (May, 1974), 158–64.

34. Cf. Lendy, *Treatise on Fortification*, p. 377.

35. Ralston B. Lattimore, *Fort Pulaski National Monument, Georgia*, National Park Service Historical Handbook Series, No. 18, p. 27.

36. Quincy A. Gillmore, *Official Report to the United States Engineer Department, on the Siege and Reduction of Fort Pulaski, Georgia*, Papers on Practical Engineering No. 8, pp. 35–36.

37. Even Quincy Gillmore, who directed the siege, was amazed at the power of the rifles. Shortly after the siege he reported, "Had we possessed our present knowledge of their power, previous to the bombardment of Fort Pulaski, the eight weeks of laborious preparation for its reduction, could have been curtailed to one week, as heavy mortars and columbiads would have been omitted from the armament of the batteries as unsuitable for breaching at long range." Gillmore, *Official Report*, p. 51.

38. Viktor Ernst Karl Rudolf von Scheliha, *A Treatise on Coast Defence: Based on the Experience Gained by Officers of the Corps of Engineers of the Army of the Confederate States...*, p. 15.

39. *Ibid.*, p. 18.

40. *The Photographic History of the Civil War*, ed. Francis Trevelyan Miller, vol. 5, *Forts and Artillery*, ed. O. E. Hunt (New York, 1911), pp. 102–4.

41. von Scheliha, *Treatise on Coast Defence*, p. 7.

42. *Ibid.*, p. 29.

43. *Ibid.*, p. 47. This principle, Number IV, was a quotation from Rear Admiral David D. Porter.

44. Among the popular articles were "The Defense of our Seaports," *Harper's New Monthly Magazine*, 71 (November, 1885), 928–37; W. B. Franklin, "National Defense," *North American Review*, 138 (December, 1883), 594–604.

45. "Report of the Board on Fortifications or Other Defenses Appointed

by the President of the United States Under the Provisions of the Act of Congress Approved March 3, 1885," 49th Cong., 1st sess., 1886, House Exec. Doc. No. 49.

46. These guns were mounted behind thick concrete parapets. They remained in a protected position below the top of the parapet until ready to fire, at which time the carriage elevated the barrels above the parapet.

4. LAND FRONTIER FORTS

1. See Stotz, "Defense in the Wilderness," pp. 185–86.

2. See J. Seymour Currey, *Chicago: Its History and Its Builders*, 5 vols. (Chicago, 1912), vol. 1, p. 58. John Whistler, a native of Ireland, served in the British army in America during the Revolution. Following the war, he enlisted in the American army and took part in western Indian campaigns. After commanding at Fort Dearborn, he was placed in charge at Fort Wayne. One of John Whistler's sons, George Washington Whistler (1800–49), became an engineer. Another son of the captain was James McNeill Whistler, widely known for his portrait of the captain's wife, commonly called "Whistler's Mother."

3. Jacob Rhett Motte, *Journey into Wilderness: An Army Surgeon's Account of Life in Camp and Field during the Creek and Seminole Wars 1836–1838*, pp. 6, 107.

4. Woodburne Potter, *The War in Florida: Being an Exposition of Its Causes, and an Accurate History of the Campaigns of Generals Clinch, Gaines, and Scott*, p. 98.

5. Motte, *Journey into Wilderness*, pp. 99, 105, 116.

6. *Ibid.*, pp. 114, 176.

7. See "Report of the Secretary of War, with Plans for the Defence and Protection of the Western Frontiers of the United States, and Statements of the Number of Indians and Warriors on those Frontiers," *American State Papers: Military Affairs*, vol. 7, p. 782.

8. "Report of a General Inspection of the Military Posts of the Western Department, and Remarks Concerning the Militia of the United States," *American State Papers: Military Affairs*, vol. 4, pp. 122, 123.

9. Major General Worth, Order No. 13, Headquarters 8th and 9th Departments, San Antonio, Texas, February 14, 1849, in Oliver Knight, *Fort Worth: Outpost on the Trinity*, pp. 232–34.

10. Report of Lieutenant Starr in C. M. Conrad, *Report of the Secretary of War*, 32d Cong., 1st sess., 1851, House Exec. Doc. No. 2, p. 270.

11. Report of Lieutenant Turnley in *ibid.*, p. 281.

12. See Earl Burk Braly, "Fort Belknap of the Texas Frontier," in *West Texas Historical Association Year Book*, 30 (1954), 97.

13. M. L. Crimmins, "Fort McKavett, Texas," *Southwestern Historical Quarterly*, 38 (July, 1934), 28.

14. Joseph K. F. Mansfield, *Mansfield on the Condition of the Western Forts, 1853–54*, p. 21.

15. *Ibid.*, p. 22.

16. J. P. Dunn, *Massacres of the Mountains: A History of the Indian Wars of the Far West, 1815–1875* (New York, 1965), p. 227.

17. U.S., Department of the Army, Surgeon General's Office, *Circular No. 4: A Report on Barracks and Hospitals with Descriptions of Military Posts*, p. 245.

18. The earthwork was a square, four-bastioned fort with redans.

19. Robert M. Utley, *Fort Union National Monument, New Mexico*, National Park Service Historical Handbook Series, No. 35, p. 34.

20. See Surgeon General's Office, *Circular No. 4*, p. 228.

21. U.S., Department of the Army, Surgeon General's Office, *Circular No. 8: A Report on the Hygiene of the United States Army with Descriptions of Military Posts*, p. 561.

22. Mansfield, *Mansfield on the Condition of the Western Forts*, p. 165.

23. *Ibid.*, plate 27.

24. Fort Laramie was established on the site of two earlier civilian posts. The first of these was Fort William, Wyoming (1834), a work made of cottonwood logs. A second structure, owned by the American Fur Company, called Fort John (1841), consisted of a rectangular enclosure built of clay with blockhouses situated on diagonal corners.

25. For this plan, see LeRoy R. Hafen and Francis Marion Young, *Fort Laramie and the Pageant of the West, 1843–1890*, p. 309.

26. Surgeon General's Office, *Circular No. 4*, p. 357.

27. Surgeon General's Office, *Circular No. 8*, p. 368.

28. *Ibid.*, p. 368.

29. *Ibid.*, p. 369.

30. Surgeon General's Office, *Circular No. 4*, p. 391.

31. Dee Brown, *Fort Phil Kearny: An American Saga*, p. 56.

32. U.S., Congress, Senate, "Report of Colonel Henry B. Carrington to Major H. G. Litchfield, July 30, 1866," *Indian Operations on the Plains*, 50th Cong., 1st sess., 1887, Senate Exec. Doc. No. 33, p. 14.

33. F. M. Fessenden, "The Building of Fort Phillip Kearney the Summer of 1866," manuscript in the E. A. Brininstool Collection, University of Texas Library, Austin.

34. Drawing, Record Group 77, Miscellaneous forts file, Cartographic Branch, National Archives, Washington, D.C.

35. Fessenden, "Building of Fort Philip Kearney."

36. Thomas G. Alexander and Leonard J. Arrington, "The Utah Military Frontier, 1872–1912: Forts Cameron, Thornburg, and Duchesne," *Utah Historical Quarterly*, 32 (Fall, 1964), 337.

37. Richard H. Coolidge, *Statistical Report on the Sickness and Mortality in the Army of the United States, Compiled from the Records of the Surgeon General's Office Embracing a Period of Sixteen Years, from January, 1839, to January, 1855*, 34th Cong., 1st sess., 1856, Senate Exec. Doc. No. 96, p. 5.

_____. "Highlights in the History of Fort Logan," *Colorado Magazine*, 19 (1942), 81–91.

Prince, L. Bradford. *Old Fort Marcy, Santa Fe, New Mexico*. Santa Fe, 1912.

Quaife, Milo M. *Chicago and the Old Northwest, 1673–1835: A Study of the Evolution of the Northwestern Frontier Together with a History of Fort Dearborn*. Chicago, 1913.

Ries, Maurice. "The Mississippi Fort Called de la Boulaye," *Louisiana Historical Quarterly*, 19 (1936), 829–99.

Riley, Edward M. "Historical Fort Moultrie in Charleston Harbor," *South Carolina Historical and Genealogical Magazine*, 51 (1950), 63–74.

Rister, Carl Coke. "The Border Post of Phantom Hill," *West Texas Historical Association Year Book*, 14 (1938), 3–13.

_____. "Fort Griffin," *West Texas Historical Association Year Book*, 1 (1925), 15–24.

_____. *Fort Griffin on the Texas Frontier*. Norman, Okla., 1956.

Robertson, Frank C. *Fort Hall: Gateway to the Oregon Country*. New York, 1963.

Robinson, Willard B. "Military Architecture at Mobile Bay," *Journal of the Society of Architectural Historians*, 30 (1971), 119–39.

_____. "North American Martello Towers," *Journal of the Society of Architectural Historians*, 33 (1974), 158–64.

Rouse, Parke, Jr. "French Antique on the Chesapeake," *Arts in Virginia*, 10 (1969), 2–9.

Ruth, Kent. *Great Day in the West: Forts, Posts and Rendezvous beyond the Mississippi*. Norman, Okla., 1963.

Sanger, Donald Bridgman. *The Story of Old Fort Bliss*. El Paso, Tex., 1933.

Scobee, Barry. *Old Fort Davis*. San Antonio, Tex., 1947.

Seymour, M. "Fort Marion, at St. Augustine—Its History and Romance," *Leslie's Monthly* (1885), 681–86.

Sides, Joseph C. *Fort Brown Historical: History of Fort Brown, Texas, Border Post on the Rio Grande*. San Antonio, Tex., 1942.

Simpson, Harold B., ed. *Frontier Forts of Texas*. Waco, Tex., 1966.

Smith, Hale G. *Santa Rosa Pensacola*. Notes in Anthropology, vol. 10. Tallahassee, Fla., 1965.

Stanley, F. [Stanley Francis Louis Crocchiola]. *Fort Bascom: Comanche-Kiowa Barrier*. Pampa, Tex., 1961.

_____. *Fort Craig*. Pampa, Tex., 1963.

_____. *Fort Stanton*. Pampa, Tex., 1964.

_____. *Fort Union*. Pampa, Tex., 1961.

Stotz, Charles Morse. *The Story of Fort Ligonier*. Fort Ligonier, Pa., 1954.

Swanberg, W. A. *First Blood: The Story of Fort Sumter*. New York, 1957.

Thomas, Daniel H. "Fort Toulouse—In Tradition and Fact," *Alabama Review*, 13 (1960), 243–57.

Thomlinson, M. H. *The Garrison of Fort Bliss, 1849–1916*. El Paso, Tex., 1945.

Tilberg, Frederick. *Fort Necessity National Battlefield Site, Pennsylvania*. National Park Service Historical Handbook Series, No. 19. Washington, D.C., 1954.

Toulouse, Joseph H., and James R. Toulouse. *Pioneer Posts of Texas*. San Antonio, Tex., 1936.

Twichel, Thomas E. "Fort Logan and the Urban Frontier," *Montana: The Magazine of Western History*, 17 (1967), 44–49.

Unrau, William E. "The Story of Fort Larned," *Kansas Historical Quarterly*, 23 (1957), 257–80.

Utley, Robert M. "Fort Union and the Santa Fe Trail," *New Mexico Historical Review*, 36 (1961), 36–48.

_____. *Fort Davis National Historic Site, Texas*. National Park Service Historical Handbook Series, No. 38. Washington, D.C., 1965.

_____. *Fort Union National Monument, New Mexico*. National Park Service Historical Handbook Series, No. 35. Washington, D.C., 1962.

Voorhis, Ernest. "Historical Forts & Trading Posts of the French Regime and of the English Fur Trading Companies." Typescript in Amon Carter Museum of Western Art, 1930.

Walsh, Richard. "Fort McHenry, 1814: The Star Fort," *Maryland Historical Magazine*, 54 (1959), 296–309.

Warner, Ted J. "Frontier Defense," *New Mexico Historical Review*, 41 (1966), 5–19.

Wenhold, Lucy L. "The First Fort of San Marcos de Apalache," *Florida Historical Quarterly*, 34 (1956), 301–14.

Wentworth, John. *Early Chicago: Fort Dearborn*. Chicago, 1881.

Whiting, J. S., and Richard J. Whiting. *Forts of the State of California*. Longview, Wash., 1960.

Wilson, Emerson W. *Fort Delaware*. Institute of Delaware History and Culture Pamphlet Series, No. 4. Newark, N.J., 1957.

Wilson, Samuel, Jr. "Colonial Fortifications and Military Architecture in the Mississippi Valley." In *The French in the Mississippi Valley*, ed. John Francis McDermott. Urbana, Ill., 1965.

Wood, Henry. "Fort Union: End of the Santa Fe Trail," *Westerners Brand Book*, 3 (1947), 205–56.

Wood, William, and Ralph Henry Gabriel. *The Pageant of America*, vol. 7, *In Defense of Liberty*. New Haven, Conn., 1928.

_____ and _____. *The Pageant of America*, vol. 6, *The Winning of Freedom*. New Haven, Conn., 1927.

Works Progress Administration. *American Guide Series*. (Books on the individual states and major cities.)

Yonge, Samuel H. *The Site of Old "James Towne," 1607–1698: A Brief Historical and Topographical Sketch of the First American Metropolis.* 5th ed. Richmond, Va., 1930.

Young, Rogers W. "The Construction of Fort Pulaski," *Georgia Historical Quarterly,* 20 (1936), 41–51.

———. *Robert E. Lee and Fort Pulaski.* National Park Service Popular Study Series, History No. 11. Washington, D.C., 1947.

Secondary Sources: Biographies

Biographie universelle ancienne et moderne.... New ed. 45 vols. Paris, 1854–65.

Blomfield, Sir Reginald Theodore. *Sébastien Le Prestre de Vauban, 1633–1707.* London, 1938.

Caemmerer, H. Paul. *The Life of Pierre Charles L'Enfant, Planner of . . . the City of Washington.* Washington, D.C., 1950.

Crouse, Nellis M. *Lemoyne d'Iberville.* Ithaca, N.Y., 1954.

Dibner, Bern. *Leonardo da Vinci: Military Engineer.* New York, 1946.

Ettinger, Amos Aschback. *James Edward Oglethorpe: Imperial Idealist.* Oxford, 1936.

Fels, Marthe de. *Terre de France: Vauban.* Paris, 1932.

Gille, Bertrand. *Engineers of the Renaissance.* Cambridge, 1966.

Halévy, Daniel. *Vauban: Builder of Fortresses.* Trans. C. J. C. Street. London, 1924.

Hamlin, Talbot. *Benjamin Henry Latrobe.* New York, 1955.

Heath, Richard Ford. *Albrecht Dürer: 1471–1520.* New York, n.d.

Holmes, Jack D. L. "Some French Engineers in Spanish Louisiana." In *The French in the Mississippi Valley,* ed. John Francis McDermott. Urbana, Ill., 1965.

Jesserand, J. J. *With Americans of Past and Present Days.* New York, 1916.

Kite, Elizabeth S. *L'Enfant and Washington, 1791–1792.* Baltimore, 1929.

Lasseray, André. *Les Français sous les treize étoiles.* Paris, 1935.

Laut, Agnes C. *Cadillac: Knight of the Wilderness, Founder of Detroit, Governor of Louisiana from the Great Lakes to the Gulf.* Indianapolis, Ind., 1931.

Nichols, James L. *The Confederate Engineers.* Tuscaloosa, Ala., 1957.

———. "Confederate Engineers and the Defense of Mobile," *Alabama Review,* 12 (1959), 181–95.

Nouvelle biographie générale depuis les temps les plus reculés jusqu'à nos jours, avec les renseignements bibliographiques et l'indication des sources à consulter. 46 vols. Paris, 1852–1866.

Robinson, William M. "The Confederate Engineers," *Military Engineer,* 22 (1930), 297–305, 410–19, 512–17.

Sautai Maurice, Theodore. *L'Oeuvre de Vauban à Lille.* Paris, 1911.

Stuart, Charles Beebe. *Lives and Works of Civil and Military Engineers of America.* New York, 1871.

Index